T0093262

Machine Learning and Deep Learning in Natural Language Processing

Natural Language Processing (NLP) is a sub-field of Artificial Intelligence, linguistics, and computer science and is concerned with the generation, recognition, and understanding of human languages, both written and spoken. NLP systems examine the grammatical structure of sentences as well as the specific meanings of words, and then they utilize algorithms to extract meaning and produce results. *Machine Learning and Deep Learning in Natural Language Processing* aims at providing a review of current Neural Network techniques in the NLP field, in particular about Conversational Agents (chatbots), Text-to-Speech, management of non-literal content – like emotions, but also satirical expressions – and applications in the healthcare field.

NLP has the potential to be a disruptive technology in various healthcare fields, but so far little attention has been devoted to that goal. This book aims at providing some examples of NLP techniques that can, for example, restore speech, detect Parkinson's disease, or help psychotherapists.

This book is intended for a wide audience. Beginners will find useful chapters providing a general introduction to NLP techniques, while experienced professionals will appreciate the chapters about advanced management of emotion, empathy, and non-literal content.

Machine Learning and Deep Learning in Natural Language Processing

Edited By
Anitha S. Pillai
Hindustan Institute of Technology and Science, Chennai, India

Roberto Tedesco
Scuola universitaria professionale della Svizzera italiana (SUPSI),
Lugano-Viganello, Switzerland

CRC Press
Taylor & Francis Group
Boca Raton London New York

CRC Press is an imprint of the
Taylor & Francis Group, an **informa** business

Designed cover image: ©ShutterStock Images

First Edition published 2024
by CRC Press
2385 NW Executive Center Drive, Suite 320, Boca Raton, FL 33431

and by CRC Press
4 Park Square, Milton Park, Abingdon, Oxon, OX14 4RN

CRC Press is an imprint of Taylor & Francis Group, LLC

ISBN: 978-1-032-26463-9 (hbk)
ISBN: 978-1-032-28287-9 (pbk)
ISBN: 978-1-003-29612-6 (ebk)

DOI: 10.1201/9781003296126

Typeset in Minion
by KnowledgeWorks Global Ltd.

Contents

Preface

NATURAL LANGUAGE PROCESSING

Machine Learning and Deep Learning in Natural Language Processsing aims at providing a review of current techniques for extracting information from human language, with a special focus on paralinguistic aspects. Such techniques represent an important part of the Artificial Intelligence (AI) research field. In fact, especially after the advent of very powerful conversational agents able to simulate a human being and interact with the user in a very convincing way, AI and the historical field of Natural Language Processing almost become synonymous (think of the abilities of GPT-3-derived models; for example, ChatGPT[1]). But let's start with a brief discussion about AI.

BRIEF INTRODUCTION TO AI

AI is the ability of machines to perform tasks that would normally require human intelligence; in particular, AI focuses on three cognitive processes: learning, reasoning, and self-correction. Historically, AI methods have been divided into two broad categories: model-driven and data-driven models. The former approach is based on a model of the task to be performed, derived by human experts looking at data; then, an algorithm is devised, based on the model. Instead, in the latter approach, the model is directly computed from data. In the following, we will focus on the latter approach.

Machine Learning (ML) is a sub-field of AI and refers to data-driven methods where systems can learn on their own, from (possibly annotated) data, without much human intervention. Using ML models, computer scientists *train* a machine by feeding it large amounts of data, the so-called *datasets*. In the so-called *supervised* approach, such datasets are annotated by human experts (e.g., think of a set of speech recordings annotated with the related transcriptions), and thus the machine tries to find the correlations among input (e.g., speech recording) and output (e.g., provided transcription). Once trained, the machine is able to perform the task on new, unknown data (e.g., new speech recordings). Another popular approach is called *unsupervised*, where the machine is trained on a dataset that does not have any labels; the goal is to discover patterns or relationships in the data. Once trained, the machine is able to apply the pattern to new data; clustering is a typical application of the unsupervised approach.

A semi-supervised approach is used when there is a combination of both labelled and unlabelled data, and labelled data is less in comparison with unlabelled data. Learning problems of this type cannot use neither supervised nor unsupervised learning algorithms, and hence it is challenging.

ML, in general, requires the developer to define the set of features (useful information extracted from "raw" data) that the model will leverage. For example, in automatic speech recognition, the Mel Frequency Cepstral Coefficient (MFCC) is the set of spectral and energy characteristics, extracted from the raw input audio samples, that classic ML models employed as input information. Selecting the right set of features is one of the most complex steps when a ML system is under development, and elicited much research efforts on feature selection, aggregation, etc.

As an evolution of ML methodologies, the Deep Learning (DL) approach starts from raw data, leaving to the model the effort of discovering useful features to describe data in an efficient and effective way. Data, thus, go through a set of layers that try to extract a more and more abstract description of them. Then, the remaining parts of the model perform the required task. This approach is useful, in two ways: (1) developers do not need to choose the features, and (2) the description found by the model is usually way better than the set of pre-defined features developers employ. A Drawback of DL is the complexity of the resulting models and the need of a huge amount of data.

In theory, many ML approaches can be "augmented" with DL, but in practice the models that are becoming the de-facto standards are based on Deep Neural Networks (DNNs, but often written as NNs). A NN is (loosely) inspired by the structure and function of the human brain; it is composed of a large number of interconnected nodes (*neurons*), which are usually organized into (many, in case of DNN) layers. Many different architectures (i.e., organization structures) have been defined so far, and probably many more will follow, permitting NNs to cope with basically any data typology (numbers, text, images, audio, etc.) executing any conceivable task. DNNs proved to be so effective that often re-defined entire research fields (think of, for example, image recognition).

Natural Language Processing (NLP) is a subset of AI, linguistics, and computer science, and it is concerned with generation, recognition, and understanding of human language, both written and spoken. NLP systems examine the grammatical structure of sentences as well as the specific meanings of words, and then they utilize algorithms to extract meaning and produce results. In other words, NLP permits to understand human language so that it can accomplish various activities automatically. NLP started in the 1950s as the intersection of AI and linguistics, and at present it is a combination of various diverse fields. In terms of NLP, task methods for NLP are categorized into two types: syntax analysis and semantic analysis. Syntax analysis deals with understanding the structure of words, sentences, and documents. Some of the tasks under this category include morphological segmentation, word segmentation, Part-of-Speech (POS) tagging, and parsing. Semantics analysis, on the other hand, deals with the meanings of words, sentences, and their combination and includes named entity recognition, sentiment analysis, machine translation, question answering, etc.

NLP MULTIMODAL DATA: TEXT, SPEECH, NON-VERBAL SIGNALS FOR ANALYSIS AND SYNTHESIS

Data is available across a wide range of modalities. Language data is multimodal and is available in the form of text, speech, audio, gestures, facial expressions, nodding the head, acoustics, and so in an ideal human–machine conversational system, machines should be able to understand and interpret this multimodal language.

Words are fundamental constructs in natural language and when arranged sequentially, such as in phrases or sentences, meaning emerges. NLP operations involve processing words or sequences of words appropriately.

Identifying and extracting names of persons, places, objects, organizations, etc., from natural language text is called the Named Entity Recognition (NER) task. Humans find this identification relatively easy, as proper nouns begin with capital letters. NER plays a major role in solving many NLP problems, such as Question Answering, Summarization Systems, Information Retrieval, Machine Translation, Video Annotation, Semantic Web Search, and Bioinformatics. The Sixth Message Understanding Conference (MUC6) introduced the NER challenge, which includes recognition of entity names (people and organizations), location names, temporal expressions, and numerical expressions.

Semantics refers to the meaning being communicated, while *syntax* refers to the grammatical form of the text. Syntax is the set of rules needed to ensure a sentence is grammatically correct; semantics is how one's lexicon, grammatical structure, tone, and other elements of a sentence combine to communicate its meaning.

The meaning of a word in Natural Language can vary depending on its usage in sentences and the context of the text. Word Sense Disambiguation (WSD) is the process of interpreting the meaning of a word based on its context in a text. For example, the word "bark" can refer to either a dog's bark or the outermost layer of a tree.

Similarly, the word "rock" can mean a "stone" or a "type of music" with the precise meaning of the word being highly dependent on its context and usage in the text. Thus, WSD refers to a machine's ability to overcome the ambiguity involved in determining the meaning of a word based on its usage and context.

Historically, NLP approaches took inspiration from two very different research fields: linguistics and computer science; in particular, linguistics was adopted to provide the theoretical basis on which to develop algorithms trying to transfer the insight of the theory to practical tasks. Unfortunately, this process proved to be quite difficult, as theories were typically too abstract to be implemented as an effective algorithm. On the other hand, computer science provided plenty of approaches, from AI and Formal Languages fields. Researchers took inspiration from practically any methodology defined in such fields, with mixed results. Thus, the result was a plethora of very different approaches, often tailored on very specific tasks, that proved difficult to generalize and often not very effective.

However, in a seminal paper published in 2003, Bengio and colleagues proposed an effective language model based on DNNs (Bengio et al., 2003). Moreover, in 2011, Collobert and colleagues proved that many NLP tasks could be greatly improved adopting DNNs (Collobert et al., 2011). Since then, ML and in particular DL and DNNs emerged as fundamental tools able to significantly improve the results obtained in many NLP tasks.

One of the most difficult tasks that classical NLP methodologies struggled to cope with is the recognition of any kind of content "hidden" in the language, such as emotions, empathy, and in general any non-literal content (irony, satirical contents, etc.). As DL promises to improve on those areas, in this book we will focus on the richness of human affective interactions and dialogues (from both textual and vocal points of view). We will consider different application fields, paying particular attention to social and critical interactions and communication, and to clinics.

HOW THIS BOOK IS ORGANIZED

We organized the chapters into five parts: I. Introduction, II. Overview of Conversational Agents, III. Sentiment and Emotions, IV. Fake News and Satire, and V. Applications in Healthcare.

In Part I, the editors introduce ML, DL, and NLP and the advancement of NLP applications using these technologies.

Part II provides an overview on current methodologies for Conversational Agents and Chatbots. Chapter 2 focuses on the applications of Chatbots and Conversational Agents (CAs) where the authors have highlighted how various AI techniques have helped in the development of intelligent CAs, and they have also compared the different state-of-the-art NLP-based chatbot architectures. An architecture of an open-domain empathetic CA designed for social conversations trained in two steps is presented in Chapter 3. The agent learns the relevant high-level structures of the conversation, leveraging a mixture of unsupervised and supervised learning, and in the second step the agent is refined through supervised and reinforcement learning to learn to elicit positive sentiments in the user by selecting the most appropriate high-level aspects of the desired response.

Part III focuses on methodologies for sentiment and emotion detection, and for production of Conversational Agent output that is augmented with emotions. In Chapter 4 authors present EMOTRON the conditioned generation of emotional speech trained with a combination of a spectrogram regression loss, to enforce synthesis, and an emotional classification style loss, to induce the conditioning.

Part IV presents methodologies for coping with fake news and satirical texts. In Chapter 5, how DL can be trained to effectively distinguish satirical content from no satire is highlighted. In Chapter 6 the authors present the development of a prototype to assist journalists with their fact-checking activities by retrieving passages from news articles that may provide evidence for supporting or rebutting the claims.

Finally, Part V shows some implementations of CA in the field of healthcare. Chapter 7 focuses on the structure and development of the algorithmic components of VocalHUM, a smart system aiming to enhance the intelligibility of patients' whispered speech in real time, based on audio to minimize the muscular and respiratory effort necessary to achieve adequate voice intelligibility and the physical movements required to speak at a normal intensity. Chapter 8 identifies the features essential for early detection of Parkinson's disease using a ML approach, and Chapter 9 explains how CAs, NLP, and ML help in psychotherapy.

NOTE

1. https://openai.com/blog/chatgpt/

REFERENCES

Yoshua Bengio, Réjean Ducharme, Pascal Vincent, and Christian Jauvin, A Neural Probabilistic Language Model, *Journal of Machine Learning Research*, 3(2003), pp. 1137–1155.

Ronan Collobert, Jason Weston, Léon Bottou, Michael Karlen, Koray Kavukcuoglu, and Pavel Kuksa, Natural Language Processing (Almost) from Scratch, *Journal of Machine Learning Research*, 12(2011), pp. 2493–2537.

Editors

Anitha S. Pillai is a professor in the School of Computing Sciences, Hindustan Institute of Technology and Science, India. She earned a Ph.D. in Natural Language Processing and has three decades of teaching and research experience. She has authored and co-authored several papers in national and international conferences and journals. She is also the co-founder of AtINeu – Artificial Intelligence in Neurology – focusing on the applications of AI in neurological disorders.

Roberto Tedesco earned a Ph.D. in Computer Science in 2006 at Politecnico di Milano in Milan, Italy, where he was contract professor for the Natural Language Processing and the Accessibility courses. He is now researcher at the Scuola universitaria professionale della Svizzera italiana (SUPSI) in Lugano, Switzerland. His research interests are NLP, assistive technologies, and HCI.

Contributors

Stefano Agresti
Politecnico Di Milano
Milan, Italy

Alice Albanesi
Scuola universitaria professionale
  della Svizzera italiana (SUPSI)
Lugano-Viganello, Switzerland

Mark J. Carman
Politecnico Di Milano
Milan, Italy

Sonia Cenceschi
Scuola universitaria professionale
  della Svizzera italiana (SUPSI)
Lugano-Viganello, Switzerland

Elisa Colletti
Scuola universitaria professionale
  della Svizzera italiana (SUPSI)
Lugano-Viganello, Switzerland

Francesco Roberto Dani
Scuola universitaria professionale
  della Svizzera italiana (SUPSI)
Lugano-Viganello, Switzerland

Naived George Eapen
Christ University
Pune, India

Claudio Ferrante
Politecnico Di Milano
Milan, Italy

Anna Giovannacci
Politecnico Di Milano
Milan, Italy

Alwin Joseph
Christ University
Pune, India

Bindu Menon
Apollo Specialty Hospitals
Nellore, India

Cristian Regna
Politecnico Di Milano
Milan, Italy

Licia Sbattella
Politecnico Di Milano
Milan, Italy

Vincenzo Scotti
Politecnico Di Milano
Milan, Italy

Alexander Sukhov
Politecnico Di Milano
Milan, Italy

Alessandro Trivilini
Scuola universitaria professionale
    della Svizzera italiana (SUPSI)
Lugano-Viganello, Switzerland

I

Introduction

Introduction to Machine Learning, Deep Learning, and Natural Language Processing

Anitha S. Pillai

Hindustan Institute of Technology and Science, Tamil Nadu, India

Roberto Tedesco

Scuola universitaria professionale della Svizzera italiana (SUPSI), Lugano-Viganello, Switzerland

1.1 ARTIFICIAL INTELLIGENCE FOR NATURAL LANGUAGE PROCESSING

Natural Language Processing (NLP) is a sub-field of computer science, information engineering, and Artificial Intelligence (AI) that deals with the computational processing and comprehension of human languages. NLP started in the 1950s as the intersection of AI and linguistics, and at present it is a combination of various diverse fields (Nadkarni et al., 2011, Otter et al., 2021). Ample volume of text is generated daily by various social media platforms and web applications making it difficult to process and discover the knowledge or information hidden in it, especially within the given time limits. This paved the way for automation using AI techniques and tools to analyze and extract information from documents, trying to emulate what human beings are capable of doing with a limited volume of text data. Moreover, NLP also aims to teach machines to interact with

DOI: 10.1201/9781003296126-2

human beings using natural language, allowing for advanced user interfaces that can be based on text or even speech.

NLP tasks can be categorized into two types: syntax analysis and semantic analysis. Syntax analysis deals with understanding the structure of words, sentences, and documents. Some of the tasks under this category include morphological segmentation, word segmentation, Part-of-Speech (POS) tagging, and parsing (Chai & Li, 2019). Semantic analysis, on the other hand, deals with the meaning of words, sentences, and their combination and includes Named Entity Recognition (NER), Sentiment Analysis, Machine Translation, Question Answering, etc. (Chai & Li, 2019).

1.2 MACHINE LEARNING AND DEEP LEARNING FOR NLP

Machine Learning (ML) for NLP and text analytics involves a set of statistical techniques for identifying parts of speech, named entities, sentiments, emotions, and other aspects of text. ML is a subset of AI which deals with the study of algorithms and statistical methods that computer systems use to effectively perform a specific task. ML does this without using explicit instructions, relying on patterns and learns from the dataset to make predictions or decisions. ML algorithms are classified into supervised, semi-supervised, active learning, reinforcement, and unsupervised learning (Langley, 1986).

1.2.1 NLP Multimodal Data: Text, Speech, and Non-Verbal Signals for Analysis and Synthesis

Data is available across a wide range of modalities. Language data is multimodal and is available in the form of text, speech, audio, gestures, facial expressions, nodding the head, and acoustics. So in an ideal human–machine conversational system, machines should be able to understand and interpret this multimodal language (Poria et al., 2021).

1.2.2 The Fundamental Role of Words

Words are fundamental constructs in natural language and when arranged sequentially, such as in phrases or sentences, meaning emerges. NLP operations involve processing words or sequences of words appropriately.

Words do not come in random order. They obey grammatical rules and convey some meaning. This characteristic of language can be captured and represented by means of a Language Model (LM). A LM, in its simplest form, is just a discrete probability distribution that permits to assign a probability value to any sequence of words:

$$P(w_1, w_2, ..., w_n)$$

Such probability distribution is usually too bit to be explicitly represented (consider that each w_i assumes values in a set that contains the whole vocabulary of the language being represented by the model), and thus several approximations are adopted (n-grams being the most popular). LMs are often the basic tool adopted in several NLP tasks.

1.2.3 Named Entity Recognition

Identifying and extracting names of persons, places, objects, organizations, etc., from natural language text is called a Named Entity Recognition (NER) task. Humans find this identification relatively easy, as proper nouns begin with capital letters. NER plays a major role in solving many NLP problems such as Question Answering, Summarization Systems, Information Retrieval, Machine Translation, Video Annotation, Semantic Web Search, and Bioinformatics. The Sixth Message Understanding Conference (MUC6) (Sundheim, 1995) introduced the NER challenge, which includes recognition of entity names (people and organizations), location names, temporal expressions, and numerical expressions.

1.2.4 Syntax and Semantics

Semantics refers to the meaning being communicated, while syntax refers to the grammatical form of the text. Syntax is the set of rules needed to ensure a sentence is grammatically correct; semantics is how one's lexicon, grammatical structure, tone, and other elements of a sentence combine to communicate its meaning.

1.2.5 Word Sense Disambiguation and Coreference Resolution

The meaning of a word in Natural Language can vary depending on its usage in sentences and the context of the text. Word Sense Disambiguation is the process of interpreting the meaning of a word based on its context in a text. For example, the word "bark" can refer to either a dog's bark or the outermost layer of a tree.

Similarly, the word "rock" can mean a stone or a type of music; the precise meaning of the word is highly dependent on its context and usage in the text. Thus, Word Sense Disambiguation refers to a machine's ability to overcome the ambiguity of determining the meaning of a word based on its usage and context.

Coreference resolution, instead, deals with finding all expressions that refer to the same entity in a text; in particular, an important sub-task is pronominal reference resolution, which deals with finding the expression a given pronouns refers to. Coreference resolution permits to improve the accuracy of several NLP tasks, such as document summarization, question answering, and information extraction.

1.3 SPEECH PROCESSING (ANALYSIS AND GENERATION)

1.3.1 Automatic Speech Recognition

Automatic Speech Recognition (ASR) refers to the task of recognizing what human beings speak and translating it into text. This research field has gained a lot of momentum over the last decades. It also plays an important role for human-to-machine communication. The earlier methods used manual feature extraction and conventional techniques such as Gaussian Mixture Models (GMMs), the Dynamic Time Warping (DTW) algorithm, and Hidden Markov Models (HMMs). Recently, neural networks – such as Recurrent Neural Networks (RNNs), Long Short-Term Memory (LSTM), Convolutional Neural Networks (CNNs), and transformers architectures, which leverage the attention mechanism, like Bidirectional Encoder Representations from Transformers (BERTs) (Huang et al., 2021) – revolutionized the field, obtaining much better results with no need for per-speaker fine-tuning.

1.3.2 Text-to-Speech

The goal of a Text-to-Speech (TTS) system is to convert text into speech. There have been numerous approaches over the years, the most prominent of which are concatenation synthesis and parametric synthesis. TTS systems take input as text and provide output as an acoustical form. ML and Deep Learning (DL) have contributed to advances in TTS, as AI-based techniques can leverage a large scale of <text, speech> pairs to learn effective feature representations to bridge the gap between text and speech and better characterize the properties of events (Ning et al., 2019).

The most natural way of human communication is through voice. It describes linguistic content (the so-called segmental level) as well as para-linguistic emotions and features (the suprasegmental level). The speaker's intended message is represented by the linguistic content, while speech prosody and other acoustic features convey paralinguistic characteristics, giving a far richer array of information about a speaker's identity, gender, age, emotions, etc. (Latif et al., 2021). HMMs and GMMs have been used extensively in speech processing, and currently they are being replaced by DL models. DL models have become an essential component in TTS, ASR, and other speech-processing tasks. The three major components of speech processing are: pre-processing, feature extraction, and ML algorithms (Latif et al., 2021).

1.3.3 Jointly Processing Text and Speech in Interactions and Communication

Some NLP tasks require (or provide better results) when text and speech are jointly processed. For example, it is well known that emotions affect both the segmental and the suprasegmental levels of human language; in other words, emotion is usually conveyed by means of specific terms (for example, the words "sad" or "happy)" and by means of subtle variation in the prosodic characteristic of voice (for example, different pitch contour or different energy contour). Thus, a model could try to leverage both "channels," working on text and speech at the same time.

Let's name E a candidate emotion, T the input text, S the input speech, E_T the decision of a model based on text, and E_S the decision of a model based on speech; we can use two models:

$$E_T = \text{argmax}_E \, P\big(E \mid T\big)$$

$$E_S = \text{argmax}_E \, P\big(E \mid S\big)$$

And then let's put together the two decisions by means of a third model that provided the final decision E_F:

$$E_F = \text{argmax}_E \, P\big(E \mid E_T, E_S\big)$$

We can train separately the first two models, one for text and one for speech, and then connect their output by means of a third, "merging" model (again, trained separately). The two initial models are easy to train, as they are coping with just one input "channel"; moreover, there is no need for a big dataset[1] containing speech and text at the same time (each model needs its own dataset). The "merging" model, which needs a text–speech dataset, can be simple as most of the work is already done by the two initial models (and thus the text–speech dataset does not need to be huge).

PATHOSnet (Scotti et al., 2020) is an example of such "ensemble" approach, where two neural networks were trained separately and then "connected" to form the final model, which was further refined.

However, this approach only partially considers the relationship between text and speech, as these two channels are processed separately. So, let's instead consider a single model that tries to handle text and speech at the same time:

$$E_F = argmax_E\, P\big(E \mid T,S\big)$$

This approach is truly able to discover and leverage the reciprocal relationships between segmental and suprasegmental levels, but it is much more complex than the method presented above. In fact, the model needs to deal with two different input typologies, and thus its structure must be more complex; moreover, training such a model requires a huge text–speech dataset.

Guo et al. (2022) is an example of such an approach where authors propose an Implicitly Aligned Multimodal Transformer Fusion (IA-MMTF) framework based on acoustic features and text information. This model enables the text and speech modalities to guide and complement each other when learning emotional representations.

1.4 AI DATA-DRIVEN METHODS AND MODELS FOR NLP

Recently introduced pre-trained LMs have the ability to address the issue of data scarcity and bring considerable benefits by generating contextualized word embeddings. These models are considered counterparts of ImageNet in NLP and have demonstrated to capture different facets of language such as hierarchical relations, long-term dependency, and sentiment (Zaib et al., 2020).

1.4.1 Introduction to Neural Data-Driven Models vs. Stochastic Data-Driven Models

A stochastic model represents a situation where there is uncertainty and can be used to model a process that has some kind of randomness. The word "stochastic" comes from the Greek word *stokhazesthai*, meaning "aim" or "guess." Uncertainty is a part of everyday life, so a stochastic model could literally represent anything. A stochastic model uses a mathematical model to derive all possible outcomes of a given problem using random input variables and focuses on the probability distribution of possible outcomes. Examples are Monte Carlo Simulation, Regression Models, and Markov Chain Models. The opposite is a deterministic model, which predicts outcomes with 100% certainty.

A neural network is a representative model of the way the human brain processes information and works by simulating a large number of interconnected processing units that resemble abstract versions of neurons. A neural network has three components: an input layer, with units representing the input fields; one or more hidden layers; and an output layer, with a unit representing the output/target fields. The units are connected with varying weights. Input data is fed to the first layer, i.e., input layer, and values propagate from each neuron to every neuron in the next layer. Eventually, a result is obtained from the output layer.

Initially, all weights are chosen at random, and the output obtained from the net is most likely incorrect. Network learns with the help of training, where examples with known outcomes are regularly presented to the network, and the responses it provides are compared to the known results (by means of a predefined loss function); if the results are not satisfactory, the weights are modified by means of a procedure called backpropagation. As training advances, the network's ability to predict and classify known outcomes improves. Once trained, the network may be applied to future situations with unknown outcomes.

1.4.2 Natural Language Processing

Machine Learning approaches such as naïve Bayes, k-nearest neighbors, HMMs, Conditional Random Fields (CRFs), decision trees, random forests, and support vector machines were popular in the past (Otter et al., 2021). However, in recent years, there has been a wholesale transformation, with neural models completely replacing, or at least augmenting, these approaches. In particular, deep neural networks, which are composed of several layers, proved to outperform the other approaches and nowadays they represent the state of the art. Note that NLP requires the models to deal with input sequences of unknown size, where the components (i.e., the words) are "interconnected" in complex ways, and often exhibit long-range dependencies; this constrain often lead to the design of approximated models that actually did not consider the whole sequence and ultimately were not able to fully capture word dependencies. Neural structures, such as RNNs (in particular, LSTMs or GRUs) or the currently best approach, called "attention mechanism", permit to handle input sequences and capture long-range word dependencies. In particular, neural networks permit to represent LMs in a very effective way, permitting to design the so-called Large Language Models (LLMs) that are then leveraged, as basic building blocks, by several network models.

1.4.3 Future Directions for Deep Learning: Explainability and Bias

Conventional NLP systems have been mostly based on techniques that are innately explainable. But over the recent years, with the increase in use of various methods and techniques like DL models, the quality has improved at the expense of being less explainable. As AI and NLP become more and more important, and users rely (sometimes blindly) on what is generated by such models, we need to ensure such models do provide correct results; thus, investigating the "behavior" of a model (or, as it is usually called, the "explainability" of a model) is becoming crucial. A well-known example is given by the so-called "hallucination" problem that affects conversational agents based on complex LLMs, where the model tends to generate content that appears to be correct but is actually false or simply senseless. This has given rise to a novel field known as Explainable AI (XAI). XAI is vital for an organization in building trust and confidence when putting AI models into production. The advantage of using XAI is providing transparency on how decisions are made by AI systems, and this in turn promotes trust between humans and machines. It can be used in healthcare, marketing, financial service, insurance, etc. XAI is expected to help in understanding why the AI system made a particular prediction or decision, or why it did not do something else.

Bias can impact ML systems at every stage, and the concept that is closely associated with bias is "fairness." An NLP system is considered to be "fair" when its outcomes are not discriminatory according to certain attributes, like gender or nationality (Garrido-Muñoz et al., 2021). Unbiased training data is an essential requirement if the deductions reached by NLP algorithms are to be trusted. According to Garrido-Muñoz et al., "The stereotyping bias in a language model is the undesired variation of the probability distribution of certain words in that language model according to certain prior words in a given domain" (Garrido-Muñoz et al., 2021).

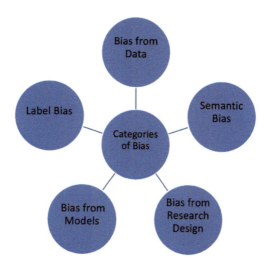

FIGURE 1.1   Different categories of bias. (Hovy & Prabhumoye, 2021.)

Bias from data arises from the dataset used for training (see Figure 1.1). For example, if we select an audio dataset from a specific demographic group, then it would be dominated by the dialect of a specific group and hence would have difficulty understanding other dialects. Annotators can also introduce label bias when they are distracted, uninterested, or lazy about the annotation task (Hovy & Prabhumoye, 2021). A balanced, well-labelled dataset also contains semantic bias: the most common text inputs represented in NLP systems, word embeddings (Mikolov et al., 2013), have been shown to pick up on racial and gender biases in the training data (Bolukbasi et al., 2016; Manzini et al., 2019). Using "better" training data alone cannot prevent bias from models, as languages evolve continuously; so even a representative sample can only capture a snapshot – at best, a temporary solution (Fromreide et al., 2014). Systems trained on biased data amplify the bias when applied to new data, and sentiment analysis tools detect societal prejudices, which results in different outcomes for various demographic groups (Zhao et al, 2017; Kiritchenko & Mohammad, 2018). Bias from research design arises as most of the work related to NLP is still conducted in and on the English language. It focuses on Indo-European data and text sources rather than other language groups or less-spoken languages, such as those found in Asia or Africa; and though there is immense data available in other languages, most NLP tools have a bias towards English (Joshi et al., 2020; Munro, 2013; Schnoebelen, 2013).

NOTE

1. Often, in the NLP field, datasets are called *corpora*; in the rest of the book, such terms will be used interchangeably.

REFERENCES

Beth M. Sundheim. Overview of results of the muc-6 evaluation. In Proceedings of the 6th conference on message understanding, MUC6 '95, pages 13–31, Stroudsburg, PA, 1995. Association for Computational Linguistics. ISBN 1-55860-402-2. doi: 10.3115/1072399.1072402. http://dx. doi.org/10.3115/1072399.1072402.

Bolukbasi, T., Chang, K.-W., Zou, J. Y., Saligrama, V., & Kalai, A. T. (2016). Man is to computer programmer as woman is to homemaker? Debiasing word embeddings. In *30th International Conference on Neural Information Processing Systems (NIPS)*.

Chai, J., & Li, A. (2019). Deep learning in natural language processing: A state-of-the-art survey. In *2019 International Conference on Machine Learning and Cybernetics (ICMLC)*. https://doi.org/10.1109/icmlc48188.2019.8949185

Garrido-Muñoz, I., Montejo-Ráez, A., Martínez-Santiago, F., & Ureña-López, L. A. (2021). A survey on bias in DEEP NLP. Applied Sciences, 11(7), 3184, pp. 1–26. https://doi.org/10.3390/app11073184

Guo, L. et al. (2022). Emotion recognition with multimodal transformer fusion framework based on acoustic and lexical information. IEEE MultiMedia, 29(02), pp. 94–103. doi: 10.1109/MMUL.2022.3161411

Hege Fromreide, Dirk Hovy, and Anders Søgaard. (2014). Crowdsourcing and annotating NER for Twitter #drift. In *Proceedings of the Ninth International Conference on Language Resources and Evaluation (LREC'14)*, pages 2544–2547, Reykjavik, Iceland. European Language Resources Association (ELRA).

Hovy, D., & Prabhumoye, S. (2021). Five sources of bias in natural language processing. Language and Linguistics Compass, 15(8). https://doi.org/10.1111/lnc3.12432

Huang, W.-C., Wu, C.-H., Luo, S.-B., Chen, K.-Y., Wang, H.-M., & Toda, T. (2021). Speech recognition by simply fine-tuning Bert. In ICASSP 2021 – *2021 IEEE International Conference on Acoustics, Speech and Signal Processing (ICASSP)*. https://doi.org/10.1109/icassp39728.2021.9413668

Joshi, P., Santy, S., Budhiraja, A., Bali, K., & Choudhury, M. (2020). The state and fate of linguistic diversity and inclusion in the NLP world. *Proceedings of the 58th Annual Meeting of the Association for Computational Linguistics*, 6282–6293. Online: Association for Computational Linguistics. https://www.aclweb.org/anthology/2020.acl-main.560

Kiritchenko, S., & Mohammad, S. (2018). Examining gender and race bias in two hundred sentiment analysis systems. in *Proceedings of the Seventh Joint Conference on Lexical and Computational Semantics*, 43–53.

Langley, P. (1986). Editorial: Human and machine learning - machine learning. SpringerLink. Retrieved July 1, 2022, from https://link.springer.com/article/10.1023/A:1022854429410

Latif, S., Qadir, J., Qayyum, A., Usama, M., & Younis, S. (2021). Speech technology for healthcare: Opportunities, challenges, and state of the art. IEEE Reviews in Biomedical Engineering, 14, 342–356. https://doi.org/10.1109/rbme.2020.3006860

Manzini, T., Yao Chong, L., Black, A. W., & Tsvetkov, Y. (2019). Black is to criminal as caucasian is to police: Detecting and removing multiclass bias in word embeddings. in *Proceedings of the 2019 Conference of the North American Chapter of the Association for Computational Linguistics: Human Language Technologies, Volume 1 (Long and Short Papers)*, 615–621. Minneapolis, MN: Association for Computational Linguistics. https://www.aclweb.org/anthology/N19-1062

Mikolov, T., Sutskever, I., Chen, K., Corrado, G. S., & Dean, J. (2013). Distributed representations of words and phrases and their compositionality. Advances in Neural Information Processing Systems, 26, 3111–3119.

Munro, R. (2013, May 22). NLP for all languages. Idibon Blog. http://idibon.com/nlp-for-all

Ning, Y., He, S., Wu, Z., Xing, C., & Zhang, L.-J. (2019). A review of deep learning based speech synthesis. Applied Sciences, 9(19), 4050. https://doi.org/10.3390/app9194050

Otter, D. W., Medina, J. R., & Kalita, J. K. (2021). A survey of the usages of deep learning for natural language processing. IEEE Transactions on Neural Networks and Learning Systems, 32(2), 604–624. https://doi.org/10.1109/tnnls.2020.2979670

Poria, S., Soon, O.Y., Liu, B. et al. (2021). Affect recognition for multimodal natural language processing. Cognitive Computation, 13, 229–230. https://doi.org/10.1007/s12559-020-09738-0

Nadkarni, P.M., Ohno-Machado, L., Chapman, W. W. (September 2011). Natural language processing: An introduction. Journal of the American Medical Informatics Association, 18(5), Pages 544–551, https://doi.org/10.1136/amiajnl-2011-000464

Schnoebelen, T. (June, 2013). *The weirdest languages*. Idibon Blog. http://idibon.com/the-weirdest-languages

Scotti, V., Galati, F., Sbattella, L., and Tedesco, R. (2020). Combining deep and unsupervised features for multilingual speech emotion recognition. International Workshop on pattern recognition for positive Technology And Elderly Wellbeing (CARE). https://link.springer.com/chapter/10.1007/978-3-030-68790-8_10

Zaib, M., Sheng, Q. Z., & Emma Zhang, W. (2020). A short survey of pre-trained language models for conversational AI-A New Age in NLP. *Proceedings of the Australasian Computer Science Week Multiconference.* https://doi.org/10.1145/3373017.3373028

Zhao, J., Wang, T., Yatskar, M., Ordonez, V., & Chang, K.-W. (2017). Men also like shopping: Reducing gender bias amplification using corpus-level constraints. *Proceedings of the 2017 Conference on Empirical Methods in Natural Language Processing*, 2979–2989. Copenhagen, Denmark: Association for Computational Linguistics.

II

Overview of Conversational Agents

CHAPTER 2

Conversational Agents and Chatbots

Current Trends

Alwin Joseph and Naived George Eapen

Department of Data Science, CHRIST (Deemed to be University), Pune Lavasa Campus, India

2.1 INTRODUCTION

Conversational agents and chatbots play an important role in daily life. Computer system conversational agents are used in creating computer-assisted human interaction systems in the educational, industrial, healthcare, and political domains [1]. We can see significant use of conversational agents to power customer interaction for product marketing and customer support in these domains. Intelligent agents are created and provided as software as a service application by various companies. This software will then be customized to the client's needs, where the engineers will create the application's intelligence. Most of the agents that are provided as a service use complex learning and algorithms to process the customer query and provide a result.

The intelligent bots created with the help of conversational agents are efficient in dealing with customer support. These systems are designed with the help of natural language technologies that are designed to process human inputs to computer understanding format. These systems are usually engaged in the initial discussion with the customer to gain more information about the customer's problem. Once the basic interaction is done, these agents capture the required information, and based on this the customers are directed to the right agent for their problem, making customer support a better experience. Similar applications for handling users are designed in most conversational agent-based systems [2]. Apart from this common approach of using intelligent agents, conversational agents also answer frequently asked questions. Textual communication agents and voice-based agents both are gaining popularity. Apart from all the advantages, the engagement of conversational agents has a huge drawback: failing to understand the user's emotions. They work the same with all people, all the time. They have the intelligence taught to them, and they fail to process the right information since they cannot mimic human behavior.

Chatbots and intelligent agents are trained to provide accurate information to the users. They are not able to generate new information. They learn from their past experiences in the advanced architectures of intelligent agents. However, most of these agents work based on the knowledge fed to them. Many industries and domains employ intelligent agents in various use cases. One such application is the use of chatbots in archaeological parks in Italy to provide contextual facts to visitors [3]. Many such interesting applications of intelligent agents can be identified apart from the conventional use of agents for simple needs. Voice-based applications like Alexa, Siri, and other interactive personal assistants also use conversational agents to communicate with humans. Natural Language Processing (NLP) is extensively used to create effective conversational agents in such applications.

The development of Artificial Intelligence (AI) and information-communication technologies help in the advancement of conversational agents in terms of organization and preciseness. Intelligent agents are being developed for textual- and voice-based communications. The use of facial recognition and gestures is one complicated area where the user agents need great improvement [4]. Researchers are focusing on these areas. However, the major drawback of such an approach is that it is challenging to identify an intelligent technology that helps the bots think and understand emotions. But as technology progresses, this problem of intelligent agents can be addressed with the help of AI and its derivatives.

Besides the traditional chatting or communicating agents' default functionality, new age conversational agents are focusing on user engagement. These systems primarily focus on the content-based systems' user engagement by creating impulsive behavioral patterns in users. This is attained by suggesting various items and contents for engaging users and making them spend more time on their websites and applications. To develop an application with user engagement and experience agents, a lot of user behavior data is required while the user is getting into such an application-based interaction system. Real-time interaction with and learning of user behavior are required in such applications. Such complex learning and adapting systems with the help of conversational agents are developed with the support of NLP, Machine Learning (ML), and web scraping techniques. Capturing user data and analyzing them to create a new market for the client is the new responsibility of intelligent agents.

2.2 CHATBOTS AND CONVERSATIONAL AGENTS

Chatbots and conversational agents are a dialogue system that enables human–computer interaction with the help of natural language for communication. This system efficiently understands and generates natural language content using text, voice, and gestures. The conversational agents effectively understand the user input, create data accordingly, and are used in various applications. With the help of AI and ML, the agents are more trained and can give precise outputs. Chatbots are one area where lots of improvements are incorporated with the help of AI. With the emergence of Deep Learning (DL) techniques, considerable progress is expected in chatbots [5]. Many researchers are focusing on creating chatbots and intelligent communication agents to incorporate DL techniques, where the primary focus is given to the performance of these applications. The algorithm used in this kind of agent can be self-learning, which can fine-tune the performance each time for effective results.

2.2.1 Conversational Agents

Conversational agents are sub-categories that fall under dialogue systems, which are tools that are designed to interact with humans. We can also call conversational agents virtual agents or assistants. Intelligent conversational agents are present in most domains for various use cases. These agents are used to communicate and gather information for various needs according to the use case designed for them. The inception of conversational agents eases the work of humans, where the computer systems are trained to function like humans to collect and gather required information for any use case. The widely used conversational agents can be grouped into various categories based on their nature.

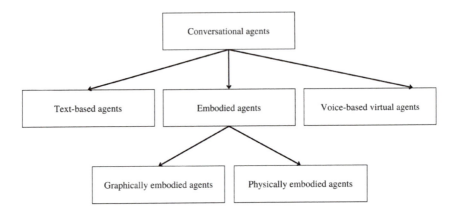

FIGURE 2.1   Types of conversational agents.

According to Figure 2.1, conversational agents help computers and humans communicate using various mediums, including text-based, voice-based, and embodied virtual agents. Embodied virtual agents are embedded into a system or hardware. Depending on how they are being embedded into some hardware components, they can be further classified into graphically embodied agents and physically embodied agents. Conversational agents are classified based on their nature [1]. Intelligent agents can be classified further into different categories based on the purpose for which they are developed [6]. However, each communication agent is designed and developed for different use cases. Mainly they are being developed for particular task-oriented and general-purpose use cases [6, 7]. Technology development has placed conversational agents as an inevitable part of daily life. Intelligent conversational agents are present in the phones we use, the customer care numbers we call for support, and many other places we encounter with other intelligent agents.

2.2.1.1 Text-Based Agents

Text-based conversational agents are systems that communicate with the users based on textual interactions. The textual agents can be called chatbots, where the user and the system chat or send messages to each other. The textual agents initially were created to collect basic information from the user that includes the name, email address, and other information depending on the use case. These kinds of textual communicable agents are common in the healthcare domain, where the bots are engaged to gather basic patient information and symptoms. Advanced and more intelligent versions of such agents are designed for patient suggestions and medications.

Text-based agents have the limitations of language. However, most are built on the English language because of ease of use and the spread and adoption of English as a global language. In most countries, English is considered a standard language; thus the systems are designed for the same language and they have been trained as well to handle the dialects of English. To achieve this adaptability, the support of NLP plays an important role. With the help of NLP, the agents can evaluate a statement, check for the data in the knowledge base, compute the decision based on the learned input, and send it to the user [8].

2.2.1.2 Voice-Based Agents

Conversational agents that focus on communication with a voice are called voice-based conversational agents. These agents are used to capture the user's sound, understand the information from the sound, analyze and perform actions accordingly, and deliver the response back to the user also in the form of voice. These agents are present in various use cases where the systems are designed to handle people's spoken words. They can be embedded into smartphones for ease of use. Such a voice-based agent is designed to obtain data for various researchers [9]. A similar voice-based agent is applied to the classroom for improving classroom engagement and the learning experience [10]. Conversational agents, especially voice-based systems, are able to identify behavioral changes [11], the stress and tension of regular users, which will help create medically intelligent bots that can assist those users. There is a wider range of applications and use cases for voice-based agents than for text-based agents. These systems are heavily dependent on the new web technologies and AI. This dependency will limit the functionality of these intelligent agents. The processing required for each individual user makes these systems a bit resource-intensive.

2.2.1.3 Embodied Agents

Embodied conversational agents are user agents with a graphical or tangible interface. The embodied agents are a combination of both text-based and voice-based agents with added features like a tangible interface and physical existence. The main concerns for the design and development of embodied agents are how the system represents the agent and its interface, how the agent gives the required information to the world and users, and how the system internally represents the interactions [12]. The best example of embodied agents is robots that are having some resilience with real-world entities like humans, animals, etc. [13]. The main sub-categories are based on how the systems have embedded the conversational agents; this includes the graphically and physically embedded conversational agents. Embodied conversational agents can handle gestures and expressions to communicate better with the users. They have wide use cases including treatments. Researchers have used such an embodied agent for communication with patients with autism using natural language and non-verbal communication [14].

2.2.2 Chatbots

Chatbots are a type of conversational agent that is text-based. The main benefit of using text-based agents, or chatbots, is that the user agent handles the queries for the users much more effectively than a human agent.

Figure 2.2 describes the different categories of chatbots based on how they are designed and developed. Rule-based chatbots are based on conditional statements set to the intelligence of the chatbots. This kind of chatbot can also be termed a "dump bot," where it will work only according to the instructions fed into the cases. If any other details are coming to the bot's concern, it will not be able to give a satisfying result to the user. Intellectually independent chatbots are based on Machine Learning (ML). The bots are built on a neural network to think and learn from similar contexts. The bot will be able to select the most relevant result for the given query. AI-powered chatbots are the most intelligent bots that can solve the most user queries. However, all bots have a knowledge base. For the rule-based chatbots, results and questions are mostly pre-trained with the expected use cases and are then deployed. On the other hand, intellectually independent and AI-powered chatbots use the knowledge base to learn and train themselves.

2.2.2.1 Rule-Based Chatbots

Rule-based chatbots are commonly known as decision-tree bots or dialogue systems [15]. These systems are based on rules defined in the knowledge base of the bots. The response of the bot is based on a series of defined

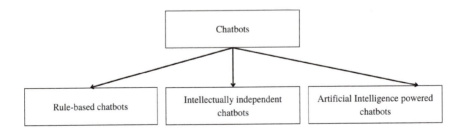

FIGURE 2.2 Types of chatbots.

rules. A lot of automated customer support bots are primarily based on rule-based bot architecture. The bots are designed and trained based on a set of rules. Once the bots are ready, they will provide the trained output to the customer based on the request. These bots are used to set up a system to convey certain messages and instructions or respond to enquiries [16]. There are many architectures for developing rule-based chatbots. Some of the prominent rule-based chatbot frameworks are Google Dialogflow and IBM Watson [17]. Rule-based chatbots are mostly text-based chatbots that normally communicate with the users via text messages and contents. The main drawback of this system is that the chatbots are not able to understand the emotions of the user. Rule-based chatbots can also be designed to be intelligent, with the help of constant learning and updating the rules and knowledge base frequently based on chat history.

2.2.2.2 Intellectually Independent Chatbots

Intellectually independent chatbots are based on ML and on the training of neural networks. Neural networks help the bots to think and learn from examples and histories. These bots are best suited for entertainment and science. They are self-learners who collect information from the related sources and train themselves to become better at serving the customers in a more efficient manner. The power of ML helps them learn about many attributes and provide a better solution for the problem at hand. The main challenge of this kind of independent bot is that it requires a person to monitor its findings and suggestions.

2.2.2.3 Artificial Intelligence-Powered Chatbots

Artificial Intelligence (AI)-powered chatbots are extended rule-based chatbots that have a mixed power of AI. These bots start with a pre-defined expected scenario that can be interrupted at any time when required. These bots use the help of Natural Language Processing (NLP) to understand the customer's text. The main benefit of this kind of bot is that it will be able to provide instant and easy information to the users. The AI-powered bots can especially take care of various people with multiple speaking abilities. These systems use the help of AI and NLP for processing the user data; with the help of the knowledge base and AI, these bots can effectively communicate with the users by understanding their emotions and circumstances in a more efficient manner than other types of chatbots. The use of NLP to process the text of users is required to understand and generate a user-specific customized experience.

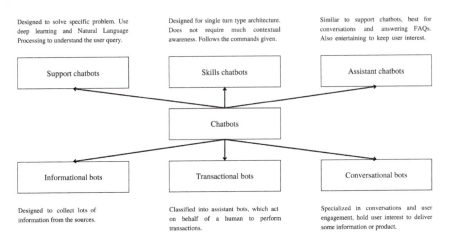

FIGURE 2.3   Classification of chatbots based on the use case.

Figure 2.3 further classifies intelligent bots into various categories based on the use cases employed. These categories help clients plan and use which type of chatbots will meet their needs. Chatbots need substantial adaption and maintenance to perform and serve the users better. The developers and maintainers of the chatbots need to carefully manage and evaluate the performance of the chatbots and constantly fine-tune their performance. There is a significant change in the perspective of users of chatbots in terms of interface design expectations and behavior [18]. Thus knowing and serving the user's expectations is always a challenging task.

2.2.3   Conversational Agents vs. Chatbots

There are a lot of similarities and differences between conversational virtual agents and chatbots. The mode of interaction and operation is the primary difference. Table 2.1 describes the differences in detail.

Aside from the differences mentioned in Table 2.1, other factors determine the most efficient tool based on which use case the systems are intended to. In most cases, conversational virtual assistants can meet a user's need, but they are expensive and require skilled personnel to manage and update. Rule-based systems, on the other hand, can be created in a much easier way and can be deployed easily. The maintenance for the chatbots can be done with the help of a simple interface where we can define and reframe rules. Both intelligent agents can communicate with humans and provide required information to them.

TABLE 2.1   Comparison of Conversational Virtual Agents and Chatbots

Conversational Virtual Agent	Chatbot
Is used for developing human–computer interaction	Communicates with computers through text
Converts human interaction into computer understanding from any form of input	Can handle only textual content
Easily understands emotions, gestures, and facial expressions	Is not able to communicate based on emotions, gestures, or facial expressions
Uses ML and NLP to understand the context and formulate responses	Behaves based on pre-defined rules
Understands the user with NLP and AI	Is not able to understand user emotions
Assists users with everyday tasks and engages in casual and fun conversations	Assists businesses and customers in their queries
Embedded in mobile phones, laptops, smart speakers, and other interactive devices	Embedded in websites, support portals, messaging channels, in-app chat widgets, and mobile applications
Accepts both textual inputs and voice-based commands	Accepts only textual inputs

2.2.4 Enhancing Customer Experience Using Chatbots

Many companies invest in and prepare chatbots to improve their customers' experience. These chatbots are virtual agents used for greeting customers initially and identifying their problems. Once the problem is identified, customers are directed to the right specialist. This helps companies to serve their customers in a more efficient manner. Besides these simple use cases, chatbots can be trained and customized for handling the following activities to support humans:

1. Lifestyle and nutritional guidance

2. Daily health checks

3. Finding a doctor

4. Medication reminder and tracking

5. Booking appointments and tickets

6. Follow-ups and remainders

7. Complaints and support

8. Food and item ordering

The main tasks that chatbots perform to enhance the customer's experience are as follows:

1. 24/7 availability

2. Seamless live chat

3. Answers for endless queries

4. Smooth interaction

5. Collection of the appropriate information

6. Automatic updates regarding queries

7. No bias against any language or culture

8. Priority to customers

Chatbots are a company's initial point of contact for almost all customer care. They gather information from customers before being directed to a human agent who can answer their queries quickly. They are essential assistants in new-age businesses used to serve the customers in an efficient manner. Customers may be apprehensive when using chatbots to assist them with a tough task. Different users have different problems, and being able to understand these problems and act accordingly is very hard for artificial bots. To enhance the user's experience and satisfaction, and ultimately their acceptance, intelligent bots should be trained with particular use cases and then deployed. But there exists a gap between real-life problems and simulated problems for training the bots.

In the case of personal assistants, user acceptance is pretty easy to achieve. Training them to the tasks is also easy, as the interaction with these agents is different. Personal assistants and similar intelligent agents are used to assist the users to remember things, personalize the technologies around them, and so on. These bots can effectively adapt to the user based on understanding the user's behavior. Thus the performance and acceptance of the intelligent bots will be efficient.

2.3 DEVELOPMENT OF CHATBOTS

Businesses and services across the globe are preparing to make their digital systems as friendly as possible to visitors. Digital systems such as websites, social media, or even applications may have to present a lot of content based on these systems' needs.

These digital systems are constantly improving through User Experience Design (UX Design). Multiple ways are introduced to present and access information stored in their servers. This can include simple ways, such as direct presentation of information in text (FAQs, etc.), or more complex methods such as conversational agents and chatbots. Irrespective of their presentation mode, the ultimate goal is to have a system that assists customers or visitors as much as possible.

Developing websites is relatively more straightforward, as it simply retrieves the corresponding content from the servers and presents it on the frontend system. However, when an intermediate agent like a search box, conversational agent, or chatbot is involved, it also needs to process the query coming in from the user's side. The fact that chatbots can interact with users in multiple ways with textual or speech skills makes the development more complex.

Different architectures are followed in the development of chatbots and conversational agents [19] and [6]. The major design frameworks are listed below:

1. Pattern Matching

2. Markov Chain Model

3. Artificial Intelligence and Machine Learning

4. Natural Language Processing

5. Database/Knowledge base

6. Web Ontology language with the help of NLP and Artificial Intelligence Mark-up Language

7. Natural Language Interaction

8. Advanced Pattern Matching and AIML

9. Artificial Neural Network Models

    a. Recurrent Neural Network

    b. Sequence to Sequence Neural Model

    c. Long Short-Term Memory Network

10. Chat Script

In all the above-mentioned frameworks, it is evident that data parsing and pre-processing are applied to the input text to generate the output by the chatbot system. This will make use of a custom-designed database and/ or knowledge base.

2.3.1 Design Process of Chatbots

When designing a conversational agent or chatbot, the designer conducts a requirement analysis to validate the design process. Some of the analysis questions include:

- What is the requirement of the chatbot (from both the business and customer perspective)?

- How is the chatbot going to be used?

- How much will data have to deal with processing the queries and replying accordingly?

Answering these questions will validate the design process of:

1. Determining the purpose of the conversational agent/chatbot

2. Choosing between a rule-based conversation agent and an NLP platform

3. Making the chatbot data-driven and training the system

4. Designing the conversation flow

5. Selecting a suitable deployment platform

Following the above process will determine whether a system should have a guided conversational agent or an NLP-based interactive chatbot. As mentioned in the earlier sections, the development of an NLP-based interactive chatbot is the hardest, as it will have to simulate a human-like interaction whether it be through a textual mode or a voice-based dialogue.

Humans communicate with each other by passing information through dialogue and conversation. When two people who know the same language speak with one another, the process of understanding and responding to each other will have a well-known structure, as depicted in Figure 2.4. A well-designed chatbot will have to follow the same process in order for effective communication to happen. Humans are able to do this process

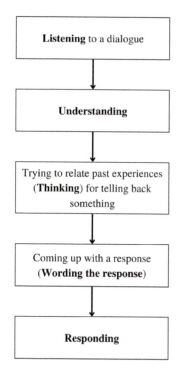

FIGURE 2.4   Process of understanding and responding.

without much effort because of the intelligence they possess. Therefore, to make machines communicate effectively, the concept of AI comes into the picture. AI has several subsets designated for specific purposes: Data Classification and Predictions, Optimizations, Big Data Processing, Image Processing, Natural Language Processing, etc.

Natural Language Processing (NLP) focuses on the problems that are extended from the process of *Understanding* and *Wording the Response*. The same is discussed in detail in the upcoming sub-sections. Conversational agents usually provide objective options for users to select from. This will involve specific paths to solutions and pre-defined question–answer sets that the users can make use of. However, these may not be able to process any problems that are out of their scope. Usually, the solution to this scenario will be to transfer the chat to a customer agent (a human) for further management of the queries. Therefore, conversational agents cannot be used as a replacement for human-led interactions. The key terms in the above process, i.e., *Listening, Understanding, Thinking, Wording the Response,* and *Responding,* become trivial because of the fixed arrangements for each solution path. While designing a chatbot, one shall comment on the approaches to solving each step in the above process. Since chatbots are likely to be used as a replacement for human agents, they will have to mimic all the steps involved in the process of understanding and responding. Thus, the process followed by humans in their conversations and data exchange is mimicked during the training process of chatbots. In other words, the training process of chatbots will ensure that the steps of Listening, Understanding, Thinking, Wording the Response, and Responding are well considered.

2.3.1.1 Listening

While communication happens, listening is the step of giving attention to what the other person tries to convey. Thinking from a chatbot or a conversational agent's point of view, the listening process points to the methods one follows to get the queries or requests from the other side, that is the user. The users generally interact with the chatbot system through a user interface or through voice-based commands. For instance, "Hey, Siri, tell me today's weather," could be taken as a voice request or a text message request.

This message will have to be transferred from the frontend side of the system to the backend side for processing and response. Sometimes, the customer support interface sends these as requests to RESTful APIs [20], such as Azure or AWS web services. It is a pure architectural or engineering problem, as it deals with sending data through multiple points of the network ensuring security until it reaches the specified endpoint. The input will have to be pre-processed and tokenized further for understanding purposes.

2.3.1.2 Understanding

Once the chatbot listens or receives the input from the user, the next process is to understand what it means. NLP enables machines to understand the input voice or text data [21]. Various technologies including linguistic-related computations, ML, or DL are used for having effective methods that can help understand even ambiguous statements. This "understanding" process could be further divided into steps of sentence segmentation, tokenization, stopword removal, adjustment related to punctuations, stemming, lemmatization, and so on [22]. These not only help in understanding the meaning but also in sentiment analysis of the sentence.

The idea of the "understanding" process is to read between the lines and capture what actually could have been meant by the user. That is, while understanding the exact sentence as read by the user, it will have to retrieve the information keeping both the context and facts that are shared. From the computer's point of view, to understand something completely, it also will have to "think" by taking the information or data that it is already familiar with. In short, the understanding process is what enables the systems to take the requested query from the user and make it ready for context-based and fact-based retrieval from the datasets by following a pre-defined method (rule-induced decisions, relating with past queries, DL methods, etc.).

2.3.1.3 Thinking

Once the system understands the query or question by performing contextual and factual analysis, it has to generate a response after a series of "thinking"-like processes. Some of the processes include taking the queries and finding useful information in the servers. That is, if the user is asking, "Give the weather status around me," it understands that the user requires information about the weather. An additional constraint to this requirement is getting the location. As the user has specified the term "around me," a series of communication happens at the server side. For instance, it will have to collect the location (latitude and longitude) of the device, followed by collecting the weather information at the same location.

This is possible because of how the system perceived the query. If any types of ambiguities are present in the query, for instance, "sum the numbers two three four and five," it will have to think whether the user asked for 234 + 5 or 2 + 3 + 4 + 5. Sometimes by using context retrieval, the thinking process can give better accuracy. Irrespective of the processes and the methodologies used, the system will have to come up with the result for the user. An important aspect here is the approach in which the chatbot engine solves the problem, whether it be a retrieval model that takes responses and gives them back as present in the system, or a generative model which actually parses meaningful sentences out of the information available in the servers.

2.3.1.4 Wording the Response

The result of the "thinking" process is the result that can be passed to the user at the frontend. Rather than presenting the answers directly, a better way is to present them in a neater sentence form. For example, when "2 + 3" is given as the question, rather than just replying "5", replying "the answer is 5" or "the sum of numbers is 5" would give a better presentation. This wording again depends on the training processes that have been done on the chatbot specifically focusing on natural language generation.

This again is a result of a rigorous training process executed at the "brain" level of the chatbot. Usually, templates are made for responding in case of very specific chatbots; however, it becomes difficult as the chatbots may have to respond to general-purpose queries as personal assistants do. The choice of the right words along with an accurate result would give a feeling that the responses of chatbots are effective.

2.3.1.5 Responding

Once the response is ready to be sent from the chatbot's brain side, it becomes a pure engineering problem of presenting to the user. In the case of a text-based chatbot, the user interface can be modified with the additional content that is received from the server (as a message bubble or as a notification, etc.). In case the chatbot is voice-based, an additional text-to-speech step also would come at the user interface's side.

The response from the server's side will have to contain the "worded response" generated from the chatbot engine's side. The same will have to be given as a JSON-based response in the case of RESTful APIs. Hence, at receiving a request from the user's side, the chatbot engine processes it, generates a suitable response, and presents it back to the user using the interfaces followed.

2.3.2 Role of Natural Language Processing

Natural Language Processing (NLP) is the core component of conversational agents. NLP has a wide variety of applications and implementations for creating conversational agents. The traditional steps of an NLP system are applicable in the generation of conversational agents as well. A detailed view is given below of various steps and processes in an NLP pipeline that is used for tracking the conversational data.

Figure 2.5 represents the role of NLP in a conversational system, where the computer system tries to capture and understand the speech of a user. The speech is captured with the help of an input source. It is pre-processed and passed into an NLP pipeline, with the help of dialogue management, and the support of a knowledge base, the appropriate results for the user are generated and provided in the form of an audio.

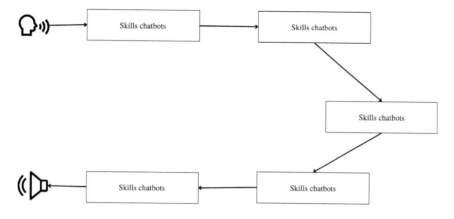

FIGURE 2.5   Role of NLP in conversational systems.

Understanding the user query is required for providing a proper response to the users. The data context is learned with the help of AI and ML techniques with the help of NLP in most of the advanced conversational agents. Conversational agents are created for various use cases some of which use NLP as their backbone including Learning and E-learning applications [23] and Healthcare applications [24]. There are a lot of ML applications and algorithms that work in the domain of NLP for effective speech recognition and processing. The popular ML techniques that are used for NLP include the following:

1. Neural Networks/Deep Learning

2. Support Vector Machines

3. Maximum Entropy

4. Bayesian Networks

5. Conditional Random Field

The major steps and segments that are part of NLP and contribute significantly to conversational agent creation include the following:

1. Natural Language Understanding (NLU)

2. Natural Language Generation (NLG)

3. Natural Language Inference (NLI)

2.3.2.1 Natural Language Understanding

Natural Language Understanding (NLU) comes under NLP, which is aimed at understanding human communication by machines. The NLU process helps the conversation agents to be precise with the user and generates results specific for the user, with the help of understanding the conversations. With NLU we can focus on learning and understanding the emotion, tone, description, and language from the users' speech data. With the help of the above factors, the conversational agent will be able to gather a basic understanding of the conversation. Apart from these common aspects, with the help of NLU, when parsing the same conversation (e.g., in e-commerce), we will be able to gather the following details from communication including organization, product, and context of the user.

The use of NLU helps the conversational agents focus on gathering the required information from the conversations. The NLU, when seen alone, performs the following tasks: relation extraction, paraphrasing/summarization, semantic parsing, sentiment analysis, and dialogue agents. These steps are processed and embedded into various phases of the development of a conversational agent for the understanding of the input from the systems.

2.3.2.2 Natural Language Generation

Natural Language Generation (NLG) is the AI technique used to generate written or spoken narratives from the data. The output that the conversational agent requires to send back to the user is passed into the NLG process to convert and generate appropriate text to the natural language. The NLG algorithms analyse the content and understand the data. After understanding the data, it is then structured, and sentence generation takes place. Once this process is completed, the grammatical structuring of the content will take place. Then the generated content will be delivered to the user. This process will remain the same for all the use cases of the NLG. Intelligent bots, to communicate with the users, require the support of the NLG process with the NLP to create statements for communication with the users. Lots of ML techniques are applied for the process of NLG [25]. NLG also helps in humanizing the experience with conversational agents [26]. Generation of NLG systems requires the expertise of language [27]. This helps in effective generation of proper content to communicate with the users.

2.3.2.3 Natural Language Inference

Natural Language Inference (NLI) is determining whether a hypothesis is True, False, or Undetermined; in other words, if the hypothesis can be inferred from a natural language "premise." If the hypothesis is True, it is called Entailment; if it is false, it is called Contradiction; and if it is Undetermined, it is called Neutral [28]. Now, understanding if the hypothesis has an entailment or contradiction, given a premise, is vital to understanding natural language, and inference about entailment and contradiction is a useful proving ground for semantic representation development. However, the absence of large-scale resources has severely hindered ML research in this domain [29].

2.3.2.4 Retrieval-Based and Generative-Based Chatbots
Different types of chatbots are designed every day based on standard datasets. Two major types that depend on the approach to wording the response are: retrieval-based chatbots and generative-based chatbots. Chatbots can converse with humans in natural language by either generating the responses or retrieving them from a set of candidate responses [30]. A comparison between these two approaches is given in Table 2.2.

TABLE 2.2   Retrieval-Based vs. Generative-Based Approaches

Retrieval-Based Dialogue Systems	Generative-Based Dialogue Systems
• Matching model • Context is encoded, and semantic matching is performed • No explicit modelling of dialogue acts, user intent, profile, etc. [30]. Different types include: single-turn matching systems (response replies to the whole context) and multi-turn matching models (response replies to each utterance of context) • Difference between these two models is in how encoders are distributed for each utterance. In single-turn matching models, only one candidate response is taken; and in multi-turn matching models, for each utterance a candidate response is taken • Single-turn matching systems include: Dual Encoders having LSTM [31], BiLSTM [32], Enhanced Sequential Inference Model [33], etc. • Multi-turn matching systems include: Sequential Matching Network [34], Deep Attention Matching Network [34, 35], Deep Utterance Aggregation system [36], etc.	• Generative probabilistic model • Phrase-based statistical machine translation • Systems training is harder due to more plausible responses and no alignment of request–response phrases [37] • Depends on a sequence-to-sequence model, where the context is stored into a summary vector • Neural conversational generative models have a kind of heuristics to make the responses based on both context and input facts [38] • Similar to how NLP does translation between languages • Instead of translating between languages, it translates from inputs to outputs • Some models that are implemented include: Task-Oriented Spoken Dialogue Systems [39], Reinforcement Learning Methods [39], Encoder-Decoder Models [40], etc.

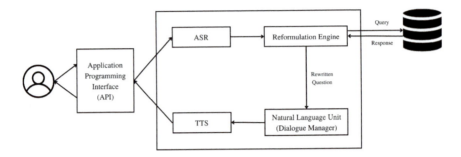

FIGURE 2.6   Architecture of a generic voice-based chatbot.

2.3.3 Case of a Voice-Based Chatbot

Considering the concepts discussed in the previous sub-sections, let us take a case of a voice-based chatbot that works for a certain application (Figure 2.6).

The use case of the above case is as follows:

1. The user tells something to the Chatbot System

2. The Chatbot System speaks out the response

Even though the use case looks very simple, it is a combined challenge of architecture, engineering, as well as NLP.

It is interesting to note the discussions happening in development communities. A detailed study [57] shows the boom in the field of ChatBot Development and how the queries are distributed in StackOverflow. Most of the queries were regarding:

- Chatbot creation and deployment

- Integration in websites and messengers

- Understanding behaviors of NLUs

- Implementation Technologies

- Development frameworks

Figure 2.7 shows the study's results regarding a chatbot topic's popularity vs. difficulty.

It is important to mention the various concepts in Figure 2.7, especially common questions raised by developers about engineering challenges related to development, such as integration of a chatbot with messengers or websites, implementation technologies, frameworks, Application Programming

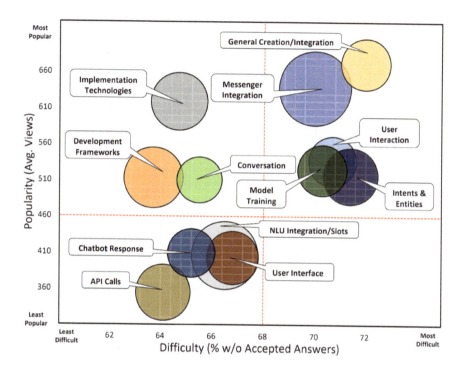

FIGURE 2.7 Chatbot topic's popularity vs. difficulty (Source: Challenges in chatbot development: A study of stack overflow posts [57]).

Interface (API) calls, and User Interface (UI) integration; whereas others are related to training of the model such as NLP processing, intents and entities, simulation of conversation, etc. All these points revolve around the development of two key parts: face (UI) and brain (trained models). Some of these terms, not discussed yet in this chapter, are briefly discussed here.

- API Calls: Application Programming Interface (API) is a software intermediary which can act as a bridge between two applications. For instance, a chatbot will have to pass the questions from the user through the user interface to the server's side using some means. This exchange happens in a request–response model, and the interaction mechanism is generally known as API calls. While making a standalone application based on customer requirements, these API calls have to be made to the server of that corresponding system. Nonetheless, developers also have a tendency to use pre-made (and easily trainable) Chatbot APIs based on a pay-per-usage basis. This will reduce the effort of the developers as we are taking the capacity of a bigger system and tuning it to the requirements of a custom system [58]. These pre-implemented functionalities could be administered to the customer needs through various microservices developed at the application's server side.

- Implementation Technologies: We have seen that Artificial Intelligence, Natural Language Processing, Machine Learning and Deep Learning are the underlying technologies for a conversational chatbot. Based on the requirements and needs of the system, one of the models as mentioned in previous subsection will be taken for development. These are implemented through programming languages such as Python or Go, and they will go through a rigorous training and testing process.

- Development Framework: A framework is a platform where one could build applications on their own, using the pre-defined settings and references the providers have made. Any frameworks would reduce the development time, effort, and cost as developers would be able to reuse the common structures which otherwise had to be written from scratch. These reliable functional models would not only speed up the process, but also ensure quality through their set rules. Some of the development frameworks related to Chatbot development are discussed in the subsequent subsection. Some of the frameworks would provide only backend technologies, whereas others would provide a combination of both backend and frontend technologies.

- User Interface: If trained models are referring to the "brain" of a system, then the User Interface becomes the "face" of the system. Normal users are not worried about the processes and transactions happening in the background. These users (or customers) would want the best experience possible, having ease of access, and good-looking and easy-to-use interfaces with minimal errors. The definition of interfaces would define how a chatbot is presented. However powerful and accurate the responses are, if the interface is not able to present the progress really well, then there is a chance that the customers may not like the product at all.

- Trained Process and Trained Models: We discussed that, currently, intent-based (or utterance-based) chatbots are the most widely used type of chatbots. Table 2.3 discusses intents and utterances; however, how these are trained is not discussed yet. Chatbots follow a supervised model of training, meaning that the input possibilities will be mapped with outputs through an iterative process. One requirement is to have a distinct set of intents to confuse the learner less. Weight adjustments happen throughout the training process like any other ML training model. The representative inputs (that is, different variations of all possible user intents) are fed into the chatbot model, and it is trained until a satisfactory convergence. Advances in the field of NLP (for instance, introduction of the BERT model [59]) have improved this training process of models to exhibit further accuracy.

TABLE 2.3   Terminologies and Challenges of the Generic High-Level Overview of a Voice-Based Chatbot

Challenge/Term	Description
Engineering Challenges	Chatbots work in a client-server model. Clients, being the users' devices, have to send a request and the server returns the apt response after a series of processes at its side. Engineering challenges denote how the front-end system makes a call, how the database takes a request and gives a result back, how user experience is enhanced, etc. Managing AI models, connecting User Interface, managing RESTful API, provisioning microservices, etc. are some common challenges [41].
Automatic Speech Recognition (ASR)	ASR helps humans use their voices to speak with a computer interface. The speech content is identified and then processed. Understanding speech and converting it to machine-understandable language is challenging and requires sophisticated speech modelling and understanding of language variations [42]. ASR is of a similar nature to pattern recognition, where the algorithm mines patterns from the speech [43]. ASR plays a significant role in speech-based chatbots and conversational agent development.
Text-to-Speech	Text-to-speech is important in converting text to sound/voice. The computer system uses this technique to speak to users, especially in voice-based conversational agents. The text required to be sent to the user is processed and organized in the system; once this is ready with the help of text-to-speech, the system can read out the required details to the user. The text-to-speech modules have a similar resemblance to humans with variations in pitch and sound [44, 45]. Such adaptive text-to-speech systems are used with the speech-based NLP models to create personalized conversational agents, but the integration of these technologies is still a challenging task.
Reformulation Engine	The reformulation engine is used to process and user query to an NLP system to extract the exact information from the user input [46]. Once the required data is captured, the algorithms search for results from the current models and databases; once the data is not captured from the data sources, the user input query will undergo a reformulation to extract the new query to be searched. This helps optimization and also supports data management [47]. The reformulation engine is used in query processing and management for extracting required data from the input and for serving users by rebuilding the application database.

(Continued)

TABLE 2.3 *(Continued)* Terminologies and Challenges of the Generic High-Level Overview of a Voice-Based Chatbot

Challenge/Term	Description
Utterance	Utterance is anything that the user says to the system. User voice inputs are considered utterances. Some examples include: "Can I have a coffee?", "How is the weather in Mumbai?", and "Can I travel to Delhi?" Some chatbots and interactive agents are designed to communicate with the users with a similar form of utterance. Some of the frameworks that support communication to the users with the utterance sentences are discussed in [48, 49, 50]. This helps the conversational agents to map to the user, easily associate with them, and meet their expectations.
Intent	Intent refers to what is behind each message that the chatbots receive. It depends on the user and the intent they communicate to the system. The intent is a critical factor that the chatbot and the conversational agent need to identify to process the message, and to give the proper response to the users, mapping to the context. Intelligent bots and agents make use of the intent to understand and replay to the users. ML algorithms are applied to identify the intent of communication; some of the algorithm implementations show that intent can be identified from the data sources effectively by ML algorithms [51, 52, 53].
Dialogue Manager	The dialogue manager is an inevitable component in the spoken dialogue system, especially in conversational agents that communicate with the users using voice. The dialogue manager has two main responsibilities: modelling and control over the dialogue flow of the system [54]. Dialogue managers help the conversational agents to keep track of the conversation and plan and organize the knowledge base for better performance and results [55]. Dialogue managers have a significant role in spoken language generation setting up and managing the context of the communication, removing ambiguity, and maintaining the flow of the communication [55, 56].

The above-mentioned case extends its scope into various learning/training methodologies. A common misconception about chatbots is that they have to pass the Turing Test. However, there are no scientific requirements for a machine to pass the Turing Test [60] to be able to "think" like humans. On the other hand, passing the Turing Test does not mean that they also possess human-like intelligence. Chatbots, like any bots, are capable of executing things that they are trained and meant to. The history of chatbots or conversational agents, starting from ELIZA to ALICE to Siri, Cortana, Google Assistant, and Alexa, all had specific training methods catering to the power and requirements at their corresponding times. Having various training methods from rule inductions or assumptions and attributions to the current scenario where data is aggressively involved in training, processes have evolved pointing to faster, efficient, and more accurate bots.

Recent advances [61] involve zealous methods of training because of the improvement in computational efficiency and advancements in DL-based models. The current state-of-the-art technologies is definitely pointing toward such DL methods. Furthermore, products are termed "personal assistants" (Alexa, Siri, Cortana, Google Assistant, etc.) because of the power that the DL algorithms could transfer to them.

2.3.4 Common APIs and Frameworks for Development

Researchers are developing various chatbots for different use cases. Some specific chatbots designed are good at what they have been developed to do. CARO – an empathetic health conversational chatbot – is an intelligent conversational agent designed to assist people who suffer from major depression [62]. It is important to note the various frameworks and application programming interfaces that are currently used by developers.

In addition to various methods for customized model training, as mentioned in the previous sub-section, there are companies that use pre-trained frameworks for reduced development time. These come in a combination of both frontend (chatbot UI) and backend (chatbot engine) models, or only the chatbot engine model.

These chatbot APIs are open to developers in free as well as paid access. While some services require their platform to be used for automatic replies, etc. (e.g., Facebook or Slack Bot APIs), other providers give their computation powers as a service (e.g., Amazon Lex) for training bots and implementing their models [63].

Some of these frameworks include:

- **Facebook Messenger API:** Facebook Messenger API provides a free platform for developing chatbots for automated-reply services on the Messenger Application. Generally, these are used for creating applications which can answer customer queries, generate posts, send messages, etc. [64].

- **Slack Bot API:** Slack, being one of the leading team-management platforms, has its own bot for answering queries from various users. The power of the pre-trained models in their default chatbot is automatically used for training various companies' bot services.

Various features, including program executions and file-based communications, are possible using the Slack Bot APIs, and they may
respond with program outputs, images, videos, files, etc. based on
contexts and requests [65].

- **Dialogflow:** Unlike the above two APIs, Dialogflow from Google is
  a platform for the management of queries from external platforms.
  That is, if one is having a standalone application or a website and
  wishes to incorporate the power of Google through chatbots, the
  Dialogflow platform can be used. The response can be in text or
  speech mode. Users are able to interact more through the virtual
  agent services (CX and ES) provided by the development API [66].

- **Amazon Lex:** Similar to Google Dialogflow, Amazon Lex is an
  Amazon Web Service (AWS) provided for building conversational
  agents and chatbots according to specific business requirements.
  By specifying the basic conversational flow in the AWS Lex con-
  sole, one is able to create bots which can provide dynamic responses
  in a conversation. It uses AWS-powered DL for Automatic Speech
  Recognition (ASR) and Spoken Language Understanding (SLU)
  problems present in the system [67].

- **Microsoft (MS) Bot Framework:** MS Bot Framework is used for
  dependable and enterprise-grade conversational AI experiences like
  Google Dialogflow or Amazon Lex. It uses the services powered by
  Azure Cognitive Services, enabling developers to connect their bots
  for virtual-assistant and customer-care operations. In addition to
  ASR, SLU, and QnA capacity, it also provides vision-based solutions
  based on the data sources through multiple channels [68].

- **Apache OpenNLP:** Apache OpenNLP is a library for processing raw
  text which can be useful for the preparation of custom bots. It supports sentence segmentation, part-of-speech tagging, tokenization,
  entity extraction, and so on. Unlike the above-mentioned frameworks, OpenNLP is a JAVA Programming Language-based library
  that can only help in the manual creation of chatbots [69].

In addition to the ones mentioned above, there are numerous paid and
open services and platforms like ChatBot API [70], PandoraBots [71], Wit
AI [72], Bot Libre [73], etc. Irrespective of the solution paths that one is
following, customers expect informative chatbots that will guide them to

come up with solutions for given queries. The power of and advancement in technology are improving research related to chatbots, particularly related to ASR, SLU, Conversational AI, Personal Assistants, etc. Following ethical principles, businesses and services would be able to come up with NLP-based solutions that can interact with customers and employees for suggestions and friendly advice as well. Imagine if an employee asks the bot, "How many leaves do I have?", and the bot replies to the query correctly. Such a system would require the retrieval of data from multiple sources (including confidential databases), good training methods, and powerful backend systems.

2.4 OTHER NOTES

Conversational agents are used in multiple sources to create communication between humans and computers. There are a lot of benefits and challenges to their usage. The use of intelligent systems is prominent in various use cases and application domains. These systems are heavily dependent on AI. The impact of such systems on human lives is beyond words. In this section, we discuss the advantages, challenges, and future scope of conversational agents.

2.4.1 Advantages

Use of chatbots and conversational agents is common in many areas. Chatbots are used to support humans in various ways. The use of intelligent agents for communication with users to collect information has significant advantages, some of which are discussed below.

2.4.1.1 Accuracy

Conversational AI systems, including chatbots, exhibit high accuracy in their assigned tasks. The systems can work more accurately than humans. Similar repeated jobs can be automated with the help of AI technologies. The use of chatbots to converse with the users to collect the relevant information for specific use cases can be effectively implemented with the help of this modern technology. The agents trained with the data can effectively convey the information to the end user without critical factors.

2.4.1.2 Efficiency

Conversational agents can effectively communicate with humans and pass the information. Users of chatbots sometimes try to divert a conversation, whether the agent is a human or an intelligent agent. However, a small gap exists between human and machine agents. Machine agents can effectively communicate with the end user without being affected by emotions. These agents work based on commands sent by humans on how to act in various situations.

2.4.1.3 Customer Experience

Customer experience of chatbot and other intelligent agent users is another important factor to consider while planning for automation. Intelligent agents sometimes fail to understand the customer's actual problem. They are programmed bots that sometimes are not able to find the best solution. However, the use of intelligent agents in various other use cases has a significant impact on the customer's perception of products and services.

2.4.1.4 Operating Costs

Operating costs of employing intelligent agents are cheaper than those for humans. They require no maintenance as far as regular operations are concerned. Also, humans cannot reach the precision and accuracy of conversational bots.

2.4.2 Challenges

The use of intelligent systems is challenging. The main challenge is data security in addition to the choice of appropriate NLP and ML techniques, understanding emotions, and conversations in native languages. These issues are described in detail below.

2.4.2.1 Data Security

Data captured by conversational agents is mostly transmitted over the internet to different locations and servers, so there exists a severe risk of data breach and security threat. Since this data is user-specific, the chance of it getting damaged is huge. Chatbot architectures must handle security, and appropriate encryption and security measures must be followed while designing applications.

2.4.2.2 Choice of Appropriate NLP and ML Techniques

The choice of appropriate NLP and ML algorithms depends on the efficiency of the conversational agents. The model required for developing the intelligence for the agents must be properly tested and their performance validated. For some use cases, generative models work effectively, while in other use cases retrieval-based models work effectively. Thus, models must be identified and properly handled in the agents' development. Inappropriate models will result in the malfunction of the system.

2.4.2.3 Understanding Emotions

Intelligent agents, based on AI and ML with the help of NLP, are trained to recognize users' emotions, but in most cases due to lack of training data, the emotion captured by the systems will not be the actual emotion of the user. Emotional understanding of the problem by the intelligent agent is a challenging task to process. This issue has led to the limited use of chatbots in many applications and cases, especially in the medical and clinical domains. Recent advancements in the field of ML, however, may lead to better understanding and optimization of the intelligent agent's performance and results.

2.4.2.4 Conversations in Native Languages

Native language conversing is another challenging task for conversational agents. A lot of chatbots concentrate on English, which is widely accepted as a global language. Very few chatbots and conversational agents concentrate on multilingualism. Thus, conversation in regional languages is a major challenge for chatbots and intelligent agents. Another major hurdle is the unavailability of the proper dataset to train the intelligent agents for multiple languages.

2.4.3 Future Scope

The development of intelligent conversational agents for communication with humans requires a lot of research and development. The systems designed to assist humans in these activities require much training and monitoring. The systems should be validated and should not harm any humans. A lot of researchers are working on different architectures for intelligent agents. The emergence of cloud computing helped in designing more intelligent and efficient systems. Most intelligent agents currently designed and developed depend on cloud-based services. Due to the cloud-based services and their dependency on them, the chatbots, communication agents, and personal assistant systems can deliver the required information to the users in milliseconds. The computations and data organizations for the same are optimized to minimize the time and maximize the performance of the systems.

Recently, serverless architectures have gained popularity and momentum in the information technology domain. Researchers are experimenting with creating chatbots with serverless computing. The serverless model used for chatbots helps extensively improve the chatbot's performance, especially for location-based content like weather reports, jokes, dates, reminders, and music [74]. There is a lot of scope for development and improvements in the intelligent agents currently used in various use cases. The future scope of research is to identify and forecast user queries and be prepared for such queries to deliver even faster results, for these intelligent architectures that are dependent on artificial intelligence, and neural networks are required.

Data security and privacy is another key concern to be addressed and handled while designing conversational bots. The data captured by the user agents if misused will be a huge threat to the person and society. These agents and assistants are with the user 24/7, so it is very difficult to keep secrets away from these systems. It should be kept in mind that attackers can easily bypass existing security measures and capture the information while designing the conversational agents. Data storage, retrieval, and management have to be optimized and organized in such a way that attackers cannot have access to people's personal data. The user agents should be tested and should use data privacy and security measures while handling the user data.

AI and DL concepts help the intelligent agents learn and understand the user queries and generate relevant results. The dependence on the knowledge base limits the agents with the designed and developed conditions. This can be improved by adding thinking intelligence to these systems. The intelligence of these agents completely depends on the data they are being trained in and the conditions they have addressed in the past. Lots of research and data are required for creating efficient bots and conversational agents. The knowledge base of the bots should also be improved with the support of various use cases.

2.5 CONCLUSION

Recent trends in chatbot design and development demonstrate the extensive use of NLP, AI, and ML for creating efficient conversational agents. There is wide use of chatbots and conversational agents as a medium of communication and engagement in various domains. Businesses have benefited from the use of chatbots for serving their customers in a more efficient way. The customer's problems are identified with the help of chatbots, and the corresponding best solutions are suggested by the intelligent agents. Humans are now much more comfortable interacting with communication agents than with other humans. It is evident that chatbots are used to collect information from patients in the healthcare domain. This shows humans' wide acceptance of chatbots and conversational agents.

Conversational agents are not effective, however, at understanding human emotions and physiological problems. But there are systems designed to handle such situations and assist and guide humans. Different architectures, such as AI and NLP, help develop chatbots that are more flexible and that better understand and communicate with humans.

The use of such complex communication agents integrating AI, ML, and NLP techniques is gaining popularity, as they can provide user-specific and appropriate communication with an understanding of user emotions. This helps businesses provide customized marketing and advertising strategies for users or a target audience. Data security, which is a common challenge in today's world, should also have some strict modes of processing and optimization to keep the user safer in the online environment.

REFERENCES

1. M. Allouch, A. Azaria, and R. Azoulay, "Conversational Agents: Goals, Technologies, Vision and Challenges," *Sensors*, vol. 21, no. 24, p. 8448, Dec. 2021 doi: 10.3390/s21248448.
2. B. Mott, J. Lester, and K. Branting, "Conversational agents," *Chapman & Hall/CRC Computer & Information Science Series*, 2004. doi: 10.1201/9780203507223.ch10.
3. Soufyane Ayanouz List Laboratory, Tangier-Morocco, Boudhir Anouar Abdelhakim List Laboratory, Tangier-Morocco, and Mohammed Benhmed List Laboratory, Tangier-Morocco, "A Smart Chatbot Architecture based NLP and Machine Learning for Health Care Assistance," *ACM Other Conferences*. https://dl.acm.org/doi/abs/10.1145/3386723.3387897 (accessed May 10, 2022).
4. M. M. E. Van Pinxteren, M. Pluymaekers, and J. G. A. Lemmink, "Human-Like Communication in Conversational Agents: a Literature Review and Research Agenda," *Journal of Service Management*, vol. 31, no. 2, pp. 203–225, Jun. 2020. doi: 10.1108/JOSM-06-2019-0175.
5. H. N. Io, and C. B. Lee, "Chatbots and conversational agents: A bibliometric analysis," *2017 IEEE International Conference on Industrial Engineering and Engineering Management (IEEM)*. 2017. doi: 10.1109/ieem.2017.8289883.
6. S. Hussain, O. Ameri Sianaki, and N. Ababneh, "A Survey on Conversational Agents/Chatbots Classification and Design Techniques," *Web, Artificial Intelligence and Network Applications*, pp. 946–956, 2019, doi: 10.1007/978-3-030-15035-8_93.
7. U. Gnewuch, S. Morana, and A. Mädche, "Towards Designing Cooperative and Social Conversational Agents for Customer Service," 2017. Accessed: May 14, 2022. [Online]. Available: https://aisel.aisnet.org/icis2017/HCI/Presentations/1

8. "Developing and evaluating conversational agents," in *Human Performance and Ergonomics*, Academic Press, 1999, pp. 173–194. doi: 10.1016/B978-012322735-5/50008-7.

9. A. S. Miner, A. Milstein, S. Schueller, R. Hegde, C. Mangurian, and E. Linos, "Smartphone-Based Conversational Agents and Responses to Questions About Mental Health, Interpersonal Violence, and Physical Health," *JAMA Internal Medicine*, vol. 176, no. 5, pp. 619–625, May 2016, doi: 10.1001/jamainternmed.2016.0400.

10. Rainer Winkler University of St. Gallen, St. Gallen, Switzerland, Sebastian Hobert University of Goettingen, Goettingen, Germany, Antti Salovaara Aalto University, Espoo, Finland, Matthias Söllner University of Kassel, Kassel, Germany, and Jan Marco Leimeister University of St. Gallen, St. Gallen, Switzerland, "Sara, the Lecturer: Improving Learning in Online Education with a Scaffolding-Based Conversational Agent," *ACM Conferences*. https://dl.acm.org/doi/abs/10.1145/3313831.3376781 (accessed May 15, 2022).

11. Rafal Kocielnik University of Washington & FXPAL, Seattle, WA, USA, Daniel Avrahami FXPAL & University of Washington, Palo Alto, CA, USA, Jennifer Marlow Google & FXPAL, Mountain View, CA, USA, Di Lu University of Pittsburgh & FXPAL, Pittsburgh, PA, USA, and Gary Hsieh University of Washington, Seattle, WA, USA, "Designing for Workplace Reflection," *ACM Conferences*. https://dl.acm.org/doi/abs/10.1145/3196709.3196784 (accessed May 15, 2022).

12. J. Cassell, "Embodied Conversational Agents: Representation and Intelligence in User Interfaces," *AIMag*, vol. 22, no. 4, pp. 67–67, Dec. 2001, doi:10.1609/aimag.v22i4.1593.

13. J. Cassell, "Embodied Conversational Interface Agents," *Communications of the ACM*, vol. 43, no. 4. pp. 70–78, 2000. doi: 10.1145/332051.332075.

14. S. Provoost, H. M. Lau, J. Ruwaard, and H. Riper, "Embodied Conversational Agents in Clinical Psychology: A Scoping Review," *Journal of Medical Internet Research*, vol. 19, no. 5. p. e151, 2017. doi: 10.2196/jmir.6553.

15. B. Liu, and C. Mei, "Lifelong Knowledge Learning in Rule-based Dialogue Systems," Nov. 2020, doi: 10.48550/arXiv.2011.09811.

16. J. Singh, M. H. Joesph, and K. B. A. Jabbar, "Rule-Based Chabot for Student Enquiries," *Journal of Physics: Conference Series*, vol. 1228, no. 1. p. 012060, 2019. doi: 10.1088/1742-6596/1228/1/012060.

17. S. A. Thorat, and V. Jadhav, "A Review on Implementation Issues of Rule-Based Chatbot Systems," *SSRN Electronic Journal*. doi: 10.2139/ssrn.3567047.

18. B. Bae and FølstadAsbjørn, "Chatbots," *Interactions*, Aug. 2018, doi: 10.1145/3236669.

19. K. Ramesh, S. Ravishankaran, A. Joshi, and K. Chandrasekaran, "A Survey of Design Techniques for Conversational Agents," *Communications in Computer and Information Science*. pp. 336–350, 2017. doi: 10.1007/978-981-10-6544-6_31.

20. C. Waghmare, "Chatbot integration," *Introducing Azure Bot Service*. pp. 111–146, 2019. doi: 10.1007/978-1-4842-4888-1_5.

21. IBM Cloud Education, "What is Natural Language Processing?" https://www.ibm.com/cloud/learn/natural-language-processing (accessed May 17, 2022).

22. J. Camacho-Collados, and M. T. Pilehvar, "On the Role of Text Preprocessing in Neural Network Architectures: An Evaluation Study on Text Categorization and Sentiment Analysis," Jul. 2017, doi: 10.48550/arXiv.1707.01780.

23. "Website." https://doi.org/10.1177/0963721414540680

24. L. Laranjo *et al.*, "Conversational Agents in Healthcare: a Systematic Review," *Journal of the American Medical Informatics Association*, vol. 25, no. 9, pp. 1248–1258, Jul. 2018, doi: 10.1093/jamia/ocy072.

25. J. Juraska, P. Karagiannis, K. K. Bowden, and M. A. Walker, "A Deep Ensemble Model with Slot Alignment for Sequence-to-Sequence Natural Language Generation," May 2018, doi: 10.48550/arXiv.1805.06553.

26. M. Virkar, V. Honmane, and S. U. Rao, "Humanizing the Chatbot with Semantics based Natural Language Generation," *2019 International Conference on Intelligent Computing and Control Systems (ICCS)*. 2019. doi: 10.1109/iccs45141.2019.9065723.

27. M.-C. Jenkins, *Designing Service-oriented Chatbot Systems Using a Construction Grammar-driven Natural Language Generation System*. 2011. [Online]. Available: https://books.google.com/books/about/Designing_Service_oriented_Chatbot_Syste.html?hl=&id=uFTuoAEACAAJ

28. C. D. Manning, and B. MacCartney, "Natural language inference," 2009. Accessed: May 17, 2022. [Online]. Available: https://www.semanticscholar.org/paper/Natural-language-inference-Manning-MacCartney/8314f8eef3b64054bfc00607507a92de92fb7c85

29. S. Bowman, G. Angeli, C. Potts, and C. D. Manning, "A large annotated corpus for learning natural language inference," in *Proceedings of the 2015 Conference on Empirical Methods in Natural Language Processing*, 2015, pp. 632–642. doi: 10.18653/v1/D15-1075.

30. B. El Amel Boussaha, N. Hernandez, C. Jacquin, and E. Morin, "Deep Retrieval-Based Dialogue Systems: A Short Review," *arXiv*, p. arXiv:1907.12878, 2019. Accessed: May 17, 2022. [Online]. Available: https://ui.adsabs.harvard.edu/abs/2019arXiv190712878E/abstract

31. S. Hochreiter, and J. Schmidhuber, "Long Short-Term Memory," *Neural Computation*, vol. 9, no. 8, pp. 1735–1780, Nov. 1997, doi: 10.1162/neco.1997.9.8.1735.

32. R. Kadlec, M. Schmid, and J. Kleindienst, "Improved Deep Learning Baselines for Ubuntu Corpus Dialogs," Oct. 13, 2015. Accessed: May 17, 2022. [Online]. Available: https://www.arxiv-vanity.com/papers/1510.03753/

33. Q. Chen, X.-D. Zhu, Z. Ling, S. Wei, and H. Jiang, "Enhancing and Combining Sequential and Tree LSTM for Natural Language Inference," 2016, Accessed: May 17, 2022. [Online]. Available: https://www.semanticscholar.org/paper/Enhancing-and-Combining-Sequential-and-Tree-LSTM-Chen-Zhu/162db03ef3cb50a07ff54ae4a1d4ea120e4162f2

34. Y. Wu, W. Wu, C. Xing, M. Zhou, and Z. Li, "Sequential Matching Network: A New Architecture for Multi-turn Response Selection in Retrieval-Based Chatbots," in *Proceedings of the 55th Annual Meeting of the Association for Computational Linguistics (Volume 1: Long Papers)*, 2017, pp. 496–505. doi: 10.18653/v1/P17-1046.

35. X. Zhou *et al.*, "Multi-Turn Response Selection for Chatbots with Deep Attention Matching Network," in *Proceedings of the 56th Annual Meeting of the Association for Computational Linguistics (Volume 1: Long Papers)*, 2018, pp. 1118–1127. doi: 10.18653/v1/P18-1103.

36. Z. Zhang, J. Li, P. Zhu, H. Zhao, and G. Liu, "Modeling Multi-turn Conversation with Deep Utterance Aggregation," in *Proceedings of the 27th International Conference on Computational Linguistics*, 2018, pp. 3740–3752. Accessed: May 17, 2022. [Online]. Available: https://aclanthology.org/C18-1317.pdf

37. H. Chen, X. Liu, D. Yin, and J. Tang, "A Survey on Dialogue Systems," SIGKDD Explorations Newsletter, Nov. 2017, doi: 10.1145/3166054.3166058.

38. N. Singh, and S. Bojewar, "Generative Dialogue System Using Neural Network," Apr. 2019, doi: 10.2139/ssrn.3370193.

39. B. Liu, G. Tür, D. Z. Hakkani-Tür, P. Shah, and L. Heck, "End-to-End Optimization of Task-Oriented Dialogue Model with Deep Reinforcement Learning," 2017, Accessed: May 17, 2022. [Online]. Available: https://arxiv.org/pdf/1711.10712.pdf

40. B. Liu, D. Hupkes, B. Calderone, J. Cheng, T. Zhao, and M. Corkery, "Generative Encoder-Decoder Models for Task-Oriented Spoken Dialog Systems with Chatting Capability," *DeepAI*, Jun. 26, 2017. https://deepai.org/publication/generative-encoder-decoder-models-for-task-oriented-spoken-dialog-systems-with-chatting-capability (accessed May 17, 2022).

41. G. Daniel, and J. Cabot, "The Software Challenges of Building Smart Chatbots," *2021 IEEE/ACM 43rd International Conference on Software Engineering: Companion Proceedings (ICSE-Companion)*. 2021. doi: 10.1109/icse-companion52605.2021.00138.

42. "Automatic Speech Recognition and Speech Variability: A Review," *Speech Communication*, vol. 49, no. 10–11, pp. 763–786, Oct. 2007, doi: 10.1016/j.specom.2007.02.006.

43. "Invited Paper: Automatic Speech Recognition: History, Methods and Challenges," *Pattern Recognition*, vol. 41, no. 10, pp. 2965–2979, Oct. 2008, doi: 10.1016/j.patcog.2008.05.008.

44. Y. Zhang, J. Cong, H. Xue, L. Xie, P. Zhu, and M. Bi, "VISinger: Variational Inference with Adversarial Learning for End-to-End Singing Voice Synthesis," *ICASSP 2022 - 2022 IEEE International Conference on Acoustics, Speech and Signal Processing (ICASSP)*. 2022. doi: 10.1109/icassp43922.2022.9747664.

45. M. Chen *et al.*, "AdaSpeech: Adaptive Text to Speech for Custom Voice," Mar. 2021, doi: 10.48550/arXiv.2103.00993.

46. *Analysis of Data on the Internet II – Search Experience Analysis*. Morgan Kaufmann, 2014, pp. e15–e26. doi: 10.1016/B978-0-12-416602-8.00015-7.

47. *Introduction.* Morgan Kaufmann, 2012, pp. 1–18. doi: 10.1016/B978-0-12-416044-6.00001-6.

48. J.-C. Gu, Z.-H. Ling, and Q. Liu, "Utterance-to-Utterance Interactive Matching Network for Multi-Turn Response Selection in Retrieval-Based Chatbots," *IEEE/ACM Transactions on Audio, Speech, and Language Processing*, vol. 28. pp. 369–379, 2020. doi: 10.1109/taslp.2019.2955290.

49. J.-C. Gu, Z.-H. Ling, and Q. Liu, "Interactive Matching Network for Multi-Turn Response Selection in Retrieval-Based Chatbots," *Proceedings of the 28th ACM International Conference on Information and Knowledge Management.* 2019. doi: 10.1145/3357384.3358140.

50. T. Yang, R. He, L. Wang, X. Zhao, and J. Dang, "Hierarchical Interactive Matching Network for Multi-Turn Response Selection in Retrieval-Based Chatbots," *Neural Information Processing.* pp. 24–35, 2020. doi: 10.1007/978-3-030-63830-6_3.

51. M. Y. H. Setyawan, R. M. Awangga, and S. R. Efendi, "Comparison of Multinomial Naive Bayes Algorithm and Logistic Regression for Intent Classification in Chatbot," *2018 International Conference on Applied Engineering (ICAE).* 2018. doi: 10.1109/incae.2018.8579372.

52. T. Pranckevičius, and V. Marcinkevičius, "Comparison of Naive Bayes, Random Forest, Decision Tree, Support Vector Machines, and Logistic Regression Classifiers for Text Reviews Classification," *Baltic Journal of Modern Computing*, vol. 5, no. 2. 2017. doi: 10.22364/bjmc.2017.5.2.05.

53. "Intent Detection and Slots Prompt in a Closed-Domain Chatbot." https://doi.org/10.1109/ICOSC.2019.8665635 (accessed May 17, 2022).

54. "The role of spoken dialogue in User–Environment interaction," in *Human-Centric Interfaces for Ambient Intelligence*, Academic Press, 2010, pp. 225–254. doi: 10.1016/B978-0-12-374708-2.00009-7.

55. "Recent trends towards cognitive science: From robots to humanoids," in *Cognitive Computing for Human-Robot Interaction*, Academic Press, 2021, pp. 19–49. doi: 10.1016/B978-0-323-85769-7.00012-4.

56. "Service Chatbots: A Systematic Review," *Expert Systems with Applications*, vol. 184, p. 115461, Dec. 2021, doi: 10.1016/j.eswa.2021.115461.

57. Ahmad Abdellatif Data-driven Analysis of Software (DAS) Lab Concordia University, Montreal, Canada, Diego Costa Data-driven Analysis of Software (DAS) Lab Concordia University, Montreal, Canada, Khaled Badran Data-driven Analysis of Software (DAS) Lab Concordia University, Montreal, Canada, Rabe Abdalkareem Software Analysis and Intelligence Lab (SAIL) Queen's University, Kingston, Canada, and Emad Shihab Data-driven Analysis of Software (DAS) Lab Concordia University, Montreal, Canada, "Challenges in Chatbot Development," *ACM Conferences.* https://dl.acm.org/doi/abs/10.1145/3379597.3387472 (accessed May 16, 2022).

58. J. Ofoeda, R. Boateng, and J. Effah, "Application Programming Interface (API) Research: A Review of the Past to Inform the Future," *International Journal of Enterprise Information Systems*, vol. 15, no. 3, pp. 76–95, 2019, doi: 10.4018/IJEIS.2019070105.

59. J. Devlin, M.-W. Chang, K. Lee, and K. Toutanova, "BERT: Pre-training of Deep Bidirectional Transformers for Language Understanding," Oct. 2018, doi: 10.48550/arXiv.1810.04805.

60. K. Warwick, and H. Shah, "Passing the Turing Test Does Not Mean the End of Humanity," *Cognitive. Computation*, vol. 8, pp. 409–419, 2016, doi: 10.1007/s12559-015-9372-6.

61. G. Caldarini, S. Jaf, and K. McGarry, "A Literature Survey of Recent Advances in Chatbots," *Information*, vol. 13, no. 1, p. 41, Jan. 2022, doi: 10.3390/info13010041.

62. N. H. I. Gandhinagar, R. S. I. Gandhinagar, S. S. I. Gandhinagar, and V. B. I. Gandhinagar, "CARO," *ACM Other conferences*. https://dl.acm.org/doi/abs/10.1145/3371158.3371220 (accessed May 10, 2022).

63. V. Srivastava, "15 Intelligent Chatbot APIs," *Nordic APIs*, Dec. 29, 2020. https://nordicapis.com/15-intelligent-chatbot-apis/ (accessed May 17, 2022).

64. "Messenger Platform - Documentation," *Facebook for Developers*. https://developers.facebook.com/docs/messenger-platform/ (accessed May 17, 2022).

65. Slack, "Enabling interactions with bots," *Slack API*. https://slack.com/bot-users (accessed May 17, 2022).

66. "Dialogflow," *Google Cloud*. https://cloud.google.com/dialogflow/docs (accessed May 17, 2022).

67. "What Is Amazon Lex?" https://docs.aws.amazon.com/lex/latest/dg/what-is.html (accessed May 17, 2022).

68. "Microsoft Bot Framework." https://dev.botframework.com/ (accessed May 17, 2022).

69. The Apache OpenNLP Team, "Apache OpenNLP." https://opennlp.apache.org/ (accessed May 17, 2022).

70. "Build your bot using ChatBot API," *ChatBot*. https://www.chatbot.com/docs/ (accessed May 17, 2022).

71. Pandorabots, Inc, "Pandorabots: Home." https://home.pandorabots.com/home.html (accessed May 17, 2022).

72. "Wit.ai." https://wit.ai/ (accessed May 17, 2022).

73. Paphus Solutions Inc, "Bot Libre." https://www.botlibre.com/ (accessed May 17, 2022).

74. Mengting Yan Department of Computer Science, University of Illinois, Urbana-Champaign, Paul Castro IBM Watson Research Center, Cambridge MA, Perry Cheng IBM Watson Research Center, Cambridge MA, and Vatche Ishakian IBM Watson Research Center, Cambridge MA, "Building a Chatbot with Serverless Computing," *ACM Conferences*. https://dl.acm.org/doi/abs/10.1145/3007203.3007217 (accessed May 10, 2022).

CHAPTER 3

Unsupervised Hierarchical Model for Deep Empathetic Conversational Agents

Vincenzo Scotti

DEIB, Politecnico di Milano, Milan, Italy

3.1 INTRODUCTION

Recent advances in Artificial Intelligence (AI) powered by Deep Learning (DL) significantly improved the state of the art in many fields [1]. In particular, the Seq2Seq transformer networks highly influenced Natural Language Processing (NLP) [2]. Transformers are Deep Neural Networks designed to deal with sequential data, as text streams, and to capture and exploit (long-term) sequential relations.

Such transformer networks can be easily fine-tuned into open-domain chatbots [3] and can be used for both retrieval and generative models. Retrieval-based chatbots select from a pool of available responses, while generative-based ones generate the sequence of tokens composing the response [1]. Generative models can adapt better to unforeseen situations, since their responses are not limited to the reference pool adopted by retrieval-based models.

This chapter is interested in a subset of the open-domain chatbots, called "empathetic chatbots" [4, 5]. These chatbots are designed and built according to models and principles of empathy, a fundamental mechanism for human interactions [6, 7]. Proper implementation of such mechanism

DOI: 10.1201/9781003296126-5

would be an essential step towards more human-like chatbots, narrowing the gap between humans and machines.

Our approach is based on Seq2Seq networks and aims at proposing a viable solution to empathetic chatbots. Moreover, we treat empathy as a control problem using a reinforcement learning approach. We train the chatbot to leverage a self-learnt, high-level dialogue structure for planning conversational acts that maximize the reward needed by the reinforcement learning approach.

We divide this chapter into the following sections. In Section 3.2, we briefly present the latest results on neural chatbots, and the current approaches for empathetic chatbots. In Section 3.3, we explain how we deal with empathy in our chatbots. In Section 3.4, we present the architecture and training procedure of the chatbot's underlying neural network. In Section 3.5, we explain how we evaluate our chatbot and present evaluation results. In Section 3.6, we summarize our work and provide hints about possible future works.

3.2 RELATED WORKS

In this section, we present a brief recap of the latest approaches for chatbots based on Seq2Seq neural networks, and we give an overview of current solutions for empathetic chatbots.

3.2.1 Seq2Seq Chatbots

Deep Neural Networks for sequence analysis (i.e., Seq2Seq models) enable the design of retrieval and generative chatbots with incredible capabilities [8, 9]. Retrieval chatbots rely on a corpus of possible dialogue turns to predict their responses. They do not suffer from disfluencies[1], but they lack flexibility, as they are limited to the set of available responses in the pool [10]. On the other hand, generative models are more flexible, as they can also produce plausible responses to unforeseen contexts [4, 11, 12], but they are prone to disfluencies.

There exist hybrid solutions combining the two approaches: retrieve-and-refine models generate starting from the retrieved response, used as an example [13], while multitask models, instead, have both retrieval and generative capabilities in a single architecture that are trained concurrently [3].

In the last ten years, the approach evolved from plain Seq2Seq causal models [14, 15] towards more complex hierarchical models, leveraging either continuous [16–18] or discrete hidden representations [19].

These early DL solutions were realized through recurrent neural networks. However, such networks were limited by the sequential analysis approach (which made it impossible to parallelize the computation) and the inability to manage extended contexts (due to the degradation of the hidden representation). Thus, current approaches rely on attention mechanisms and transformer architectures [2].

Due to the high availability of pre-trained transformers [20], it is now possible to fine-tune them into conversational agents without the need for long training sessions on huge corpora, while still achieving impressive results [3]. These chatbot models are able to capture complex long-term relationships (and hence longer contexts) and allow for completely parallel computation [11, 12, 21, 22].

Usually, these neural chatbots are trained with a supervised learning approach. Given the context (i.e., the considered conversation history), retrieval models are trained to maximize the posterior probability of the whole target response. Instead, generative models are trained to maximize the log-likelihood of the next token in the response, given the context and the preceding response tokens (autoregressive approach).

The training approach is not limited to the supervised one; it is possible to rely on a reinforcement learning approach, and indeed many solutions have been proposed to train open-domain agents through reinforcement learning [23–25]. Unlike task-oriented chatbots, in the case of open-domain chatbots, the reward is not well defined. Thus, metrics measuring social conversational skills and conversation goodness are used as rewards in the reinforcement learning problem. Various solutions were proposed to measure such aspects, even through learnt metrics [26].

3.2.2 Empathetic Chatbots

As premised, empathetic conversational agents are a sub-class of open-domain chatbots; in particular, such agents try to perceive emotions and react to them showing empathy, a fundamental mechanism of human–human interaction. Empathy can be roughly described as the ability to understand another's inner state and, possibly, respond accordingly (more on this in Section 3.3).

In the last few years, a growing interest in this area led to the proposal of several solutions to implement empathy in conversational agents. XiaoIce represents an impressive example of an empathetic agent [4]. It implements both emotional and cognitive aspects of empathy, and it is powered by knowledge, persona, and image grounding. Moreover, it is deployed on many social media, thus having access to user profiles for a more personalized experience (in fact, it is possible to mine useful personal information from websites like Twitter or Facebook [27]). Additionally, it is embodied through voice and an avatar, making it easier to perceive the agent as a human.

The agent embodiment through visual and voiced interaction modules, although quite powerful in improving the user experience, is hard to manage. Thus, other solutions limit the interaction to text exchanges, yielding a more straightforward development process. Conversational agents like CAiRE [5], MoEL [28], and EmpTransfo [29] implement empathy in their ability to recognize the user's emotion or predict the most appropriate response emotion, dialogue act, and more. These agents learn the emotional mechanism, generating conditioned text; in other words, they have the ability to generate a response given some high-level attributes, like the desired emotion.

Such textual models learn to simulate empathy by imitating good empathetic behavior examples, but it is possible to go beyond this approach: setting an explicit objective compatible with an empathetic behavior makes it possible to implicitly train the agent towards an empathetic behavior.

The idea is to set an objective that implicitly requires the agent to understand the user's inner state from the conversation context, and to act accordingly. In particular, the agents' goal is to elicit a positive sentiment in the user. Agents like Emo-HRED [30] or MC-HRED [31] select the desired high-level response attributes (emotion and dialogue act) to maximize such target. Thus, the actual response generation is conditioned on the selected attributes. This approach is also directly applicable at a lower dialogue level, like in the sentiment look-ahead network [32]. This network leverages reinforcement learning to alter the probability distribution of the next token in the response, maximizing the positivity of the user's expected sentiment.

3.3 APPROACH TO EMPATHY

Empathetic computing is a generalization of affective computing [33]. Early works on affective computing explained how a machine would not be completely intelligent as long as it does not perceive the user's emotions. Empathy completes this concept by explaining how it is important to show emotional and cognitive intelligence [6]. These aspects of empathy allow understanding someone's mental state (like emotion or intent).

Empathy affects human interactions at different levels, as in a hierarchy, and multiple frameworks reflect that [6, 7]. In this work, we propose following this same approach and building the chatbot to see the conversation from a hierarchical perspective. The idea is to treat empathy as a control problem and have the agent selecting a high-level abstract response first, and then yielding the low-level response, all according to an empathetic policy. This policy controls the empathetic behavior of the agent [34].

We relied on data-driven approaches to build our generative empathetic chatbot, following the impressive advances observed in open-domain agents [4, 11–13, 21]. Our approach started from a probabilistic language model; in particular, we used a pre-trained language model to have strong initialization in features and generative capabilities, and we fine-tuned it into the final dialogue language model.

Unlike previous works on empathetic agents, we propose leveraging unsupervised learning to extract a discrete high-level dialogue model during the dialogue language modelling training. Previous works rely on high-level labels (like emotion or dialogue act) available on annotated corpus [5, 28, 29]. This may represent a limitation since not all corpora are annotated or are based on the same label set. Our approach allows merging multiple corpora and thus training a more complex model with possibly better generative capabilities.

We further fine-tuned the agent using reinforcement learning on an empathetic objective to provide the agent with empathetic capabilities. We refined the agent to maximize the user's positive sentiment (extracted from the next conversation turn) and the user's engagement (measured as the next turn relative length, with respect to the previous one). This step is necessary for the agent to learn the aforementioned empathetic policy. We use a discrete high-level model and a hybrid training framework to ensure this refinement step does not break the agent's conversational capabilities [19].

3.4 CHATBOT IMPLEMENTATION

In this section, we describe the probabilistic language model we used to implement our dialogue agent, and the training process we followed to embed the agent with empathy[2].

3.4.1 DLDLM Architecture

As premised, we build our agent through a probabilistic dialogue language model; in particular, we designed and implemented it starting from the well-known GPT-2 [35] language model. Then, we extended the resulting model to include the hierarchical aspects of language we want to learn and exploit.

We extended the vanilla Seq2Seq architecture of GPT-2 with additional heads (i.e., final linear transformations). The idea was to learn a set of discrete latent codes by clustering the responses while learning to predict them. In doing that, we follow an approach similar to PLATO [21, 22]; however, our approach also predicts latent codes, while PLATO only uses posterior recognition.

The resulting dialogue language model is a variational auto-encoder with discrete latent codes. The model learns the latent codes in an unsupervised way and uses the recognized or predicted latent codes (at train and inference time, respectively) to condition the response generation. We call this architecture Discrete Latent Dialogue Language Model (DLDLM).

The model takes the sequence of context tokens x_c and the sequence of response tokens as input. The response tokens can be either those of the correct response x_r or a distractor x_d (for multiobjective training; more on this in Section 3.4.2). During training, the model's input comprises the entire sequence of response tokens. At inference time, instead, the response tokens are generated in an autoregressive fashion. The overall input structure is presented in Figure 3.1.

The model fetches three kinds of embeddings that sum together at each position in the sequences, to encode the input and feed the hidden transformations. We distinguish among token embeddings, token type embeddings, and position embeddings.

Token embeddings are the regular embeddings calculated from the textual sequence. We wrap each turn with special token embeddings to indicate the beginning (<s>) and end (<s/>) of each of them. We also introduce additional embeddings to encode the latent codes z. Finally, we have

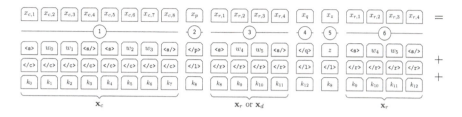

FIGURE 3.1 Input structure: Top-row elements are the actual input embeddings to the hidden transformation, second-row elements are token embeddings, third-row elements are token type embeddings, and fourth-row elements are position embeddings. The wi tokens were identified by the original GPT-2 tokenizer. Numbers in circles identify the steps in input processing.

special tokens to instruct the model to perform the posterior (</q>) or prior (</p>) latent analysis.

There are three types of token embeddings that represent where the tokens come from: context (</c>), response (</r>), or latent analysis (</l>). Finally, we use position embeddings to encode positional information into the token representation.

In addition to the hidden transformations $h(\cdot)$, the model has six distinct heads:

- We use the Language Modelling head $y_{lm}(x_c, z, x_r)$ to predict the probability of the next response token: $P(x_{r,i} \mid x_c, z, x_{r,j<i})$;

- We use the Latent Posterior head $y_q(x_c, x_r)$ to predict the posterior latent distribution $P(z|x_c, x_r) \overset{\text{def}}{=} Q$;

- We use the Latent Prior head (or Policy head) $y_p(x_c)$ to predict the prior latent distribution $P(z|x_c) \overset{\text{def}}{=} P$;

- We use the Classification head $y_{cls}(x_c, x_r)$ to predict the posterior probability that a given response is correct $P(C = \text{correct} \mid x_c, x_r)$, and also the posterior probability that a given distractor response is wrong $P(C = \text{wrong}|x_c, x_d) = 1 - P(C = \text{correct} \mid x_c, x_d)$;

- We use the Bag-of-Words (BoWs) head $y_{bow}(x_c, z)$ to predict the normalized BoW representation of the response $\text{BoW}(x_r) = y_{bow}(x_c, z)$;

- We use the Reward head $\hat{r} = y_{rew}(x_c, z)$ to predict the immediate reward r.

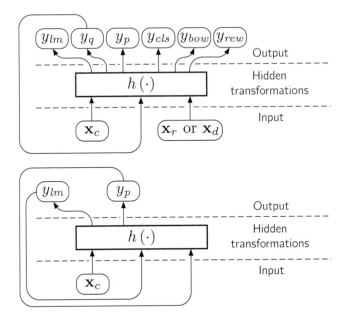

FIGURE 3.2 Model abstract architecture and input/output flow. (a) Training. (b) Inference.

The heads are used differently, depending on whether the model is deployed at train or inference time, as depicted in Figure 3.2 (more on this in Section 3.4.2).

Following the number notation in Figure 3.1, the model follows this pipeline:

1. The model encodes x_c into the encoded context $H_c = h(x_c)$ using the hidden transformations.

2. The model encodes the prior latent analysis token </p> into $h_p = h(x_p, H_c)$, given the encoded context H_c, and predicts the prior probability distribution P using $y_p(\cdot)$.

3. The model encodes x_r into the encoded response $H_r = h(x_c, H_c)$, given the encoded context H_c. During training, x_d is encoded too, as an alternative path to x_r.

4. The model encodes the posterior latent analysis token </q> into $h_q = h(x_q, H_c, H_r)$, given the encoded context H_c and response H_r, and predicts the posterior probability distribution Q using $y_q(\cdot)$.

Then, the model computes the posterior probability of a response to be the correct one (for both x_r and x_d), on top of h_q, using the retrieval head $y_{cls}(\cdot)$.

5. The model encodes the selected high-level latent token z into $h_z = h(z, H_c)$ from the encoded context H_c. During training, z is sampled from Q; during inference, from P. Then, on top of h_z, the model predicts the expected reward with $y_{rew}(\cdot)$, and predicts the BoW representation with $y_{bow}(\cdot)$.

6. Finally, the model computes the posterior probability of the next token $x_{r,i}$, given the encoded context H_c, the encoded latent h_z, and the preceding response tokens $x_{r,j<i}$, using the language modelling head $y_{lm}(\cdot)$.

3.4.2 Training

We train the DLDLM model in two steps.

During the first step, the model learns the high- and low-level dialogue model. We leveraged unsupervised learning to extract the high-level model and supervised learning to extract the low-level dialogue model.

During the second step, the model learns the empathetic behavior. We leveraged a hybrid reinforcement and supervised learning approach to learn the empathetic policy without breaking the underlying dialogue model.

3.4.2.1 Discrete Latent Dialogue Language Model

We trained the network on the first step using mini-batches X of dialogue response samples. Each sample is a quadruple composed of a sequence of context tokens x_c, a sequence of response tokens x_r, a sequence of distractor tokens x_d, and an immediate reward vector r. We update the parameters Θ of the network to minimize the loss described in the following equation:

$$\mathcal{L}(X; \Theta) = \mathbb{E}_X\left[\mathcal{L}_{LM}(x_c, x_r)\right] + \mathbb{E}_X\left[\mathcal{L}_{KLt}(x_c, x_r)\right] + \mathbb{E}_X\left[\mathcal{L}_{CLS}(x_c, x_r, x_d)\right]$$

$$+ \mathbb{E}_X\left[\mathcal{L}_{BoW}(x_c, x_r)\right] + \sqrt{\mathbb{E}_X\left[\mathcal{L}_{REW}(x_c, x_r, r)\right]}$$

where:

- $\mathcal{L}_{LM}(\cdot)$ is the average negative log-likelihood of observing the response tokens given the context tokens and the preceding response tokens (i.e., usual language modelling loss) to train the dialogue language model.

- $\mathcal{L}_{KLt}(\cdot)$ is the thresholded Kullback-Leibler (KL) divergence of $P(\cdot)$ from $Q(\cdot)$, used to prevent the vanishing KL issue [36] and train the discrete latent model [37].

- $\mathcal{L}_{CLS}(\cdot)$ is the contrastive binary cross-entropy to train the retrieval model.

- $\mathcal{L}_{BoW}(\cdot)$ is the average negative log-likelihood of the response tokens computed from z, to help training the latent model.

- $\mathcal{L}_{REW}(\cdot)$ is the mean squared reward prediction error, to help modelling the hidden features for the following training step.

During this training step, we modify the activation of the posterior head $y_q(\cdot)$. We employ a Gumbel-Softmax$(\cdot)$ [38] instead of the regular Softmax$(\cdot)$, as, during training, we are interested in dealing with a distribution as close as possible to the categorical one (due to the discrete approach), while still needing to maintain the latent sampling process differentiable.

3.4.2.2 Empathetic Policy

During the second step, we trained the network using mini-batches X of episodes E (i.e., entire dialogues). Then, for training the empathetic controller (i.e., empathetic policy) we resorted to a policy gradient algorithm: REINFORCE [39]. In particular, we used the off-policy version of the algorithm to avoid wasting resources for conversation simulations, and to avoid introducing errors due to the possible faults in the dialogue generation process (sometimes, models like the one we are designing tend to yield dull or inconsistent responses [8]).

To avoid breaking the generative capabilities learnt from the previous step, we resorted to a hybrid reinforcement and supervised training objective [32] to maximize the hybrid objective function described in the following equation. The two objectives are weighted by a parameter $\lambda \in [0,1] \subseteq \mathbb{R}$ to control the trade-off between the reinforcement learning objective $J_{RL}(E)$ and the supervised learning loss $\mathcal{L}_{SL}(\cdot)$, in the hybrid training.

$$J(E;\Theta) = \lambda \mathbb{E}\big[J_{RL}(E)\big] + (1-\lambda)\mathbb{E}\big[\mathcal{L}_{SL}(E)\big]$$

$$J_{RL}(E;\Theta) = -\sum_{t=1}^{|E|} \tilde{G}^{(t)} \cdot \Big(\mathcal{L}_{NLLz}\big(x_c^{(t)}, x_r^{(t)}\big) + \alpha \mathcal{L}_{LM}\big(x_c^{(t)}, x_r^{(t)}\big)\Big)$$

where:

- $J_{RL}(\cdot)$ is the reinforcement learning objective to maximize, computed as in the previous equation.

- $\mathcal{L}_{SL}(\cdot)$ is the supervised learning loss to minimize, defined as in the first equation, but with $\mathcal{L}_{NLLz}(\cdot)$ instead of $\mathcal{L}_{KLt}(\cdot)$ (see the next point).

- $\mathcal{L}_{NLLz}(\cdot)$ is the negative log-likelihood of predicting the latent code that maximizes $Q(\cdot)$ using $P(\cdot)$.

- $\alpha \in \{0, 1\}$ is a parameter to control whether to use the REINFORCE objective to also influence the low-level language modelling ($\alpha = 1$) or only the high-level policy ($\alpha = 0$).

- $\tilde{G}^{(t)}$ is a standardized cumulative discounted reward computed under the behaviours policy at time step t.

As in the previous step, we resorted to using the Gumbel-Softmax$(\cdot)$.

3.4.2.3 Hyperparameters

We trained and refined two versions on the network based on the 117 and 345 million parameter versions of the original GPT-2.

The two models were trained for 30 and 10 epochs, respectively, during the first training step, and for a single epoch in the second one. During the first training step, we used a mini-batch of size 64, and during the second training step we used a mini-batch of size 1 (a single episode).

In each context-response pair we considered only contexts up to 256 tokens and responses up to 128 tokens. We leveraged the original GPT-2 tokenizer to encode the turn strings.

Regarding the training process, we used the AdamW Optimizer [40], and, in all training processes, we adopted a linear learning rate schedule with 0.2% of update steps warmup. The maximum learning rates in the two implementations were $6.25 \cdot 10^{-5}$ and $3.125 \cdot 10^{-5}$, respectively.

Finally, the Gumbel-Softmax$(\cdot)$ used a temperature rescoring of $T = 2/3$.

3.5 EVALUATION

In this section, we present the approach we followed in the evaluation of the agent, the corpora employed, and the subsequent results.

3.5.1 Corpora

We trained and evaluated our chatbot on a mix of different well-curated corpora to have sufficient data to extract a reliable high-level model. In particular, we merged four different open-domain conversation corpora: DailyDialog (DD) [41], EmpatheticDialogues (ED) [42], Persona-Chat (PC) [43], and Wizard of Wikipedia (WoW) [44]. We used the same splits of the original corpora to collect the train and validation samples we used in the learning steps, and the test samples we used in the evaluation steps. Table 3.1 reports the main statistics about the corpora.

As premised, we considered two distinct rewards to maximize, in the empathetic learning step. The elicited sentiment reward was computed scaling the results of sentiment analysis of each turn, in a $[-1, 1] \subseteq \mathbb{R}$ range. The reward about the relative response length was computed as the difference between the number of next turn tokens and current response ones, normalizing on the current response length; this reward was further scaled through a $\tanh(\cdot)\$$ to constrain the values in a $[-1, 1] \subseteq \mathbb{R}$ range. We leveraged an external tool (SpaCy³ library) to compute these values.

3.5.2 Approach

We evaluated the chatbot implementations through automatic metrics to assess the quality of the dialogue language model and to assess the positive effects of the empathetic refinement. The 117M and 345M models were compared right after the first training step, after the policy fine-tuning, and after policy and language modelling joined the fine-tuning. In this way, we observed the effects of the various training steps.

To evaluate the generative capabilities of the dialogue language mode, we resorted to Perplexity (PPL) [3, 11, 12]. It is the most commonly used metric for this kind of evaluation. Moreover, it strongly correlates with human judgment on dialogue quality [12].

We maintain an off-policy approach to evaluate empathy and sociality, as in the training step [24]. Thus, we compute the average cumulative reward of the models, weighted on the probability of doing the same action under the baseline policy or the empathetic policies.

We split the evaluation in two parts to better observe the effects of fine-tuning for empathy, at different granularity levels.

TABLE 3.1 Main Statistics on the Considered Corpora Organized per Split

	Train			Validation			Test		
	Dialogues	Turns Per Dialogue	Tokens Per Turn	Dialogues	Turns Per Dialogue	Tokens Per Turn	Dialogues	Turns Per Dialogue	Tokens Per Turn
DD	11118	7.84 ± 4.01	14.37 ± 10.83	1000	8.07 ± 3.88	14.28 ± 10.52	1000	7.74 ± 3.84	14.56 ± 10.92
ED	19533	4.31 ± 0.71	15.90 ± 9.80	2770	4.36 ± 0.73	17.08 ± 9.66	2547	4.31 ± 0.73	18.16 ± 10.38
PC	8939	14.70 ± 1.74	12.11 \pm 4.24	1000	15.60 ± 1.04	12.37 ± 4.05	968	15.52 ± 1.10	12.23 ± 4.00
WoW	18430	9.05 ± 1.04	19.88 ± 9.64	981	9.08 ± 1.02	19.89 ± 9.62	965	9.03 ± 1.02	19.91 ± 9.58
Total	58020	8.09 ± 3.99	15.97 ± 9.33	5751	7.77 ± 4.45	15.49 ± 8.81	5480	7.75 ± 4.45	15.76 ± 9.15

TABLE 3.2   Results of the PPL Off-Policy Evaluation of the Language Model

Model		PPL			
Configuration	Size	All data	$r_{sent.} > 0$	$r_{soc.} > 0$	$r_{sent.} > 0 \wedge r_{soc.} > 0$
Baseline	117M	18.27 ± 31.08	18.23 ± 31.76	17.22 ± 35.48	17.36 ± 36.73
	345M	14.67 ± 21.88	14.65 ± 22.32	13.93 ± 24.87	14.03 ± 25.85
Emp. (π)	117M	18.54 ± 33.51	18.46 ± 33.99	17.41 ± 38.28	17.51 ± 39.49
	345M	14.85 ± 21.80	14.80 ± 21.88	14.04 ± 24.18	14.11 ± 24.75
Emp. (π, LM)	117M	27.65 ± 70.93	27.42 ± 72.87	24.38 ± 79.75	24.60 ± 83.33
	345M	18.85 ± 62.54	18.90 ± 67.16	16.93 ± 32.81	17.03 ± 33.47

Note: Emp. (π) models refer to the policy fine-tuning; Emp. (π, LM) models refer to the policy and language modelling joined fine-tuning; the remaining models are the baselines (i.e., no fine-tuning). $r_{sent.} > 0$ refers to the samples in the test set where the elicited sentiment reward is non-negative; $r_{soc.} > 0$ refers to the samples in the test set where the tanh($\cdot$) of the elicited response relative length is non-negative.

3.5.3 Results

We reported the results of the PPL evaluation[4] in Table 3.2 and the results of the empathetic policy (controller) in Figure 3.3. As premised, the reported results are from automatic metrics. A human evaluation should be carried out to understand the actual chatbot behavior in a better way. As for now, we limited the evaluation to this automatic approach to gathering early results on the proposed approach.

Concerning PPL, the first result we point out is that higher model complexity reflects in the results. The trained models achieved lower PPL scores when used in the 345M version. This lower PPL score reflects other results

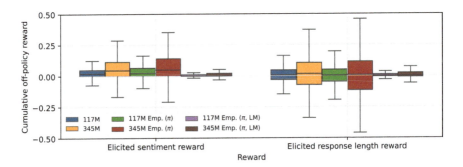

FIGURE 3.3   Results of the off-policy evaluation of the empathetic controller. Emp. (π) models refer to the policy fine-tuning; Emp. (π, LM) models refer to the policy and language modelling joined fine-tuning; the remaining models are the baselines (i.e., no fine-tuning).

in literature, where authors showed how increasing model complexity does improve language modelling capabilities [11, 12].

Another point to highlight is how empathetic fine-tuning negatively affects language modelling capabilities. The "policy only" fine-tuning does not sensitively affect the PPL; this is expected since we did not alter the language modelling loss to train the model. The "joined" fine-tuning, however, produces way worse results. The 345M model ends up with a PPL closer to the 117M model than the other two tested configurations. Despite this being expected when computing PPL on the whole corpus, since the model needs to reject responses that may have negative rewards, we expected better results when considering only the subset of interactions with positive rewards (last three columns of Table 3.2). Given the results of similar works [32], we expect that a better hyper-parameter search could help fix this issue.

Finally, we would like to point out that the model always performs better on samples associated with positive rewards: it is correctly oriented towards responses that can promote users' positive sentiment and longer responses. Although a deeper analysis is required, the results we obtained can be a hint that our approach is viable to introduce an empathetic behavior in conversational agents.

We immediately noticed two aspects of the weighted cumulative rewards value ranges concerning the empathetic policy. Empathetic fine-tuning of policy and language modelling leads to narrow ranges than other results. Moreover, the 345M models cover a more comprehensive range of values than the 117M ones.

Models that undergo fine-tuning of policy and language modelling achieve acceptable results in the evaluation of the weighted cumulative reward (as shown in Figure 3.3). The overall distribution of values is mostly non-negative, meaning that the model is oriented towards actions with non-negative rewards. This was the expected result of the fine-tuning. However, the narrow ranges indicate that instead of going toward positive rewards, the model learnt a "safe" policy where the rewards are close to 0. This behavior is a common issue of off-policy learning. This result and the low PPL scores lead us to the realization that this fine-tuning at multiple levels of granularity requires an ad-hoc analysis to work correctly, which we leave as possible future work.

Models that undergo empathetic fine-tuning only on the policy partially confirm the results from the PPL analysis. Observing the distributions of the cumulative elicited sentiment rewards, we notice that the model achieves higher maximum rewards and averages than the baseline counterparts. These higher scores mean that the empathetic fine-tuning positively affected the model towards a more empathetic behavior, favoring the user's positive sentiment. Observing the cumulative distribution of the elicited response's relative lengths, however, we do not find the same behavior: maxima are higher than the baseline counterparts, but not averages. However, most of the distribution is non-negative, showing that the fine-tuning did not lead to undesired behaviors.

Finally, we point out that models that did not undergo empathetic fine-tuning still achieved good results in this evaluation. These results are primarily due to the corpus. Despite presenting examples of responses to cover both positive and negative rewards, there is an unbalance towards positive scores; thus, the model learns this behavior directly from the training samples. Ideally, we would need a balanced corpus to have a sharper effect after fine-tuning; in practice, these data are hard to find, especially among well-curated dialogue corpora.

From these results, we evinced that the empathetic fine-tuning, limited to the high-level aspects of the conversation, achieves better results on elicited sentiment, showing a viable solution for the development of empathetic chatbots. Moreover, acting only at a high level helps not to disrupt the language modelling capabilities of the agent (the difference in PPL between these models and the baseline counterparts can be considered negligible).

3.6 CONCLUSION AND FUTURE WORK

This chapter describes our solution to implement and train an empathetic chatbot using a Seq2Seq approach. The agent is trained in a two-step process, starting from a pre-trained probabilistic language model. During the first step, we fine-tune the agent to generate dialogue and learn a discrete latent dialogue structure. In the second step, we resort to hybrid reinforcement and supervised learning to exploit the dialogue structure and the dialogue generative capabilities, further refining the agent to optimize empathy-related rewards.

In our empathetic agent, we approach empathy as a control problem. We train and evaluate different versions of the Seq2Seq neural network in the experiments. The rewards we train the agent to optimize are the elicited positive sentiment (to enforce emotional intelligence) and the relative response length (to enforce a social behavior that pushes the user towards openness). Applying the control at different levels of granularity, we observe that DLDLM produces better results when fine-tuned for empathy at the high-level dialogue model only.

As of now, we foresee two possible directions. On one side, we are willing to refine the agent on more task-oriented conversations; the idea is to keep the open-domain conversation setting but with an overall goal requiring empathy and others' understanding, like in therapy or counselling sessions. On the other side, we are interested in completing the chatbot adding modules for voiced input/output, namely an Automatic Speech Recognition and a Text-to-Speech system. These extensions would make the agent appear more human and thus more relatable, a fundamental property for empathetic agents.

NOTES

1. In this context, the term "disfluency" means the generation of meaningless sentences.
2. The code base with the dialogue agent model and the training process are available at https://github.com/vincenzo-scotti/dldlm/tree/v2.0
3. https://spacy.io
4. For the models that undergo empathetic fine-tuning on both policy and language modelling, we got many infinite PPLs; to be able to compute these values, we filtered all PPL > 10,000 considering them as outliers (more comments about this later).

REFERENCES

1. D. Jurafsky and J. H. Martin, "Speech and language processing: an introduction to natural language processing, computational linguistics, and speech recognition (3rd Edition)," https://web.stanford.edu/~jurafsky/slp3/, 2022.
2. Q. Liu, M. J. Kusner and P. Blunsom, "A Survey on Contextual Embeddings," *CoRR*, vol. abs/2003.07278, 2020.
3. T. Wolf, V. Sanh, J. Chaumond and C. Delangue, "TransferTransfo: A Transfer Learning Approach for Neural Network Based Conversational Agents," *CoRR*, vol. abs/1901.08149, 2019.
4. L. Zhou, J. Gao, D. Li and H.-Y. Shum, "The Design and Implementation of XiaoIce, an Empathetic Social Chatbot," *Comput. Linguistics*, vol. 46, pp. 53–93, 2020.
5. Z. Lin, P. Xu, G. I. Winata, F. B. Siddique, Z. Liu, J. Shin and P. Fung, "CAiRE: An End-to-End Empathetic Chatbot," in *The Thirty-Fourth AAAI Conference on Artificial Intelligence, AAAI 2020, The Thirty-Second Innovative Applications of Artificial Intelligence Conference, IAAI 2020, The Tenth AAAI Symposium on Educational Advances in Artificial Intelligence, EAAI 2020, New York, NY, USA, February 7–12, 2020.*

6. M. Asada, "Towards Artificial Empathy – How Can Artificial Empathy Follow the Developmental Pathway of Natural Empathy?" *Int. J. Soc. Robotics*, vol. 7, pp. 19–33, 2015.

7. ÖN. Yalçin, "Empathy Framework for Embodied Conversational Agents," *Cogn. Syst. Res*, vol. 59, pp. 123–132, 2020.

8. J. Gao, M. Galley and L. Li, "Neural Approaches to Conversational AI," in *Proceedings of ACL 2018, Melbourne, Australia, July 15–20, 2018, Tutorial Abstracts*, 2018.

9. M. Huang, X. Zhu and J. Gao, "Challenges in Building Intelligent Open-Domain Dialog Systems," *ACM Trans. Inf. Syst*, vol. 38, pp. 21:1–21:32, 2020.

10. X. Gao, Y. Zhang, M. Galley, C. Brockett and B. Dolan, "Dialogue Response Ranking Training with Large-Scale Human Feedback Data," in *Proceedings of the 2020 Conference on Empirical Methods in Natural Language Processing, EMNLP 2020, Online, November 16–20*, 2020.

11. Y. Zhang, S. Sun, M. Galley, Y.-C. Chen, C. Brockett, X. Gao, J. Gao, J. Liu and B. Dolan, "DIALOGPT: Large-Scale Generative Pre-training for Conversational Response Generation," in *Proceedings of the 58th Annual Meeting of the Association for Computational Linguistics: System Demonstrations, ACL 2020, Online, July 5–10, 2020*.

12. D. Adiwardana, M.-T. Luong, D. R. So, J. Hall, N. Fiedel, R. Thoppilan, Z. Yang, A. Kulshreshtha, G. Nemade, Y. Lu and Q. V. Le, "Towards a Human-Like Open-Domain Chatbot," *CoRR*, vol. abs/2001.09977, 2020.

13. S. Roller, E. Dinan, N. Goyal, D. Ju, M. Williamson, Y. Liu, J. Xu, M. Ott, E. M. Smith, Y.-L. Boureau and J. Weston, "Recipes for Building an Open-Domain Chatbot," in *Proceedings of the 16th Conference of the European Chapter of the Association for Computational Linguistics: Main Volume, EACL 2021, Online, April 19–23, 2021*.

14. A. Ritter, C. Cherry and W. B. Dolan, "Data-Driven Response Generation in Social Media," in *Proceedings of the 2011 Conference on Empirical Methods in Natural Language Processing, EMNLP 2011, 27–31 July 2011, John McIntyre Conference Centre, Edinburgh, UK, A meeting of SIGDAT, a Special Interest Group of the ACL*, 2011.

15. O. Vinyals and Q. V. Le, "A Neural Conversational Model," *CoRR*, vol. abs/1506.05869, 2015.

16. A. Sordoni, M. Galley, M. Auli, C. Brockett, Y. Ji, M. Mitchell, J.-Y. Nie, J. Gao and B. Dolan, "A Neural Network Approach to Context-Sensitive Generation of Conversational Responses," in *NAACL HLT 2015, The 2015 Conference of the North American Chapter of the Association for Computational Linguistics: Human Language Technologies, Denver, Colorado, USA, May 31–June 5, 2015*.

17. I. V. Serban, A. Sordoni, Y. Bengio, A. C. Courville and J. Pineau, "Building End-To-End Dialogue Systems Using Generative Hierarchical Neural Network Models," in *Proceedings of the Thirtieth AAAI Conference on Artificial Intelligence, February 12–17, 2016, Phoenix, Arizona, USA*, 2016.

18. I. V. Serban, A. Sordoni, R. Lowe, L. Charlin, J. Pineau, A. C. Courville and Y. Bengio, "A Hierarchical Latent Variable Encoder-Decoder Model for

Generating Dialogues," in *Proceedings of the Thirty-First AAAI Conference on Artificial Intelligence, February 4–9, 2017, San Francisco, California, USA*, 2017.

19. C. Sankar and S. Ravi, "Deep Reinforcement Learning For Modeling Chit-Chat Dialog With Discrete Attributes," in *Proceedings of the 20th Annual SIGdial Meeting on Discourse and Dialogue, SIGdial 2019, Stockholm, Sweden, September 11–13, 2019*.

20. T. Wolf, L. Debut, V. Sanh, J. Chaumond, C. Delangue, A. Moi, P. Cistac, T. Rault, R. Louf, M. Funtowicz, J. Davison, S. Shleifer, P. von Platen, C. Ma, Y. Jernite, J. Plu, C. Xu, T. L. Scao, S. Gugger, M. Drame, Q. Lhoest and A. M. Rush, "Transformers: State-of-the-Art Natural Language Processing," in *Proceedings of the 2020 Conference on Empirical Methods in Natural Language Processing: System Demonstrations, EMNLP 2020 - Demos, Online, November 16–20, 2020*.

21. S. Bao, H. He, F. Wang, H. Wu, H. Wang, W. Wu, Z. Guo, Z. Liu and X. Xu, "PLATO-2: Towards Building an Open-Domain Chatbot via Curriculum Learning," in *Findings of the Association for Computational Linguistics: ACL/IJCNLP 2021, Online Event, August 1–6, 2021*.

22. S. Bao, H. He, F. Wang, H. Wu e and H. Wang, "PLATO: Pre-trained Dialogue Generation Model with Discrete Latent Variable," in *Proceedings of the 58th Annual Meeting of the Association for Computational Linguistics, ACL 2020, Online, July 5–10, 2020*.

23. J. Li, W. Monroe, A. Ritter, D. Jurafsky, M. Galley and J. Gao, "Deep Reinforcement Learning for Dialogue Generation," in *Proceedings of the 2016 Conference on Empirical Methods in Natural Language Processing, EMNLP 2016, Austin, Texas, USA, November 1–4, 2016*.

24. I. V. Serban, C. Sankar, M. Germain, S. Zhang, Z. Lin, S. Subramanian, T. Kim, M. Pieper, S. Chandar, N. R. Ke, S. Mudumba, A. de Brébisson, J. Sotelo, D. Suhubdy, V. Michalski, A. Nguyen, J. Pineau and Y. Bengio, "A Deep Reinforcement Learning Chatbot," *CoRR*, vol. abs/1709.02349, 2017.

25. A. Saleh, N. Jaques, A. Ghandeharioun, J. H. Shen and R. W. Picard, "Hierarchical Reinforcement Learning for Open-Domain Dialog," in *The Thirty-Fourth AAAI Conference on Artificial Intelligence, AAAI 2020, The Thirty-Second Innovative Applications of Artificial Intelligence Conference, IAAI 2020, The Tenth AAAI Symposium on Educational Advances in Artificial Intelligence, EAAI 2020, New York, NY, USA, February 7–12, 2020*.

26. R. Lowe, M. Noseworthy, I. V. Serban, N. Angelard-Gontier, Y. Bengio and J. Pineau, "Towards an Automatic Turing Test: Learning to Evaluate Dialogue Responses," in *Proceedings of the 55th Annual Meeting of the Association for Computational Linguistics, ACL 2017, Vancouver, Canada, July 30 – August 4, Volume 1: Long Papers*, 2017.

27. M. Brambilla, A. J. Sabet and A. E. Sulistiawati, "Conversation Graphs in Online Social Media," in *Web Engineering – 21st International Conference, ICWE 2021, Biarritz, France, May 18–21, 2021, Proceedings*, 2021.

28. Z. Lin, A. Madotto, J. Shin, P. Xu and P. Fung, "MoEL: Mixture of Empathetic Listeners," in *Proceedings of the 2019 Conference on Empirical Methods in Natural Language Processing and the 9th International Joint Conference on Natural Language Processing, EMNLP-IJCNLP 2019, Hong Kong, China, November 3–7, 2019.*

29. R. Zandie and M. H. Mahoor, "EmpTransfo: A Multi-Head Transformer Architecture for Creating Empathetic Dialog Systems," in *Proceedings of the Thirty-Third International Florida Artificial Intelligence Research Society Conference, Originally to be held in North Miami Beach, Florida, USA, May 17–20, 2020.*

30. N. Lubis, S. Sakti, K. Yoshino and S. Nakamura, "Eliciting Positive Emotion through Affect-Sensitive Dialogue Response Generation: A Neural Network Approach," in *Proceedings of the Thirty-Second AAAI Conference on Artificial Intelligence, (AAAI-18), the 30th innovative Applications of Artificial Intelligence (IAAI-18), and the 8th AAAI Symposium on Educational Advances in Artificial Intelligence (EAAI-18), New Orleans, Louisiana, USA, February 2–7, 2018.*

31. N. Lubis, S. Sakti, K. Yoshino and S. Nakamura, "Unsupervised Counselor Dialogue Clustering for Positive Emotion Elicitation in Neural Dialogue System," in *Proceedings of the 19th Annual SIGdial Meeting on Discourse and Dialogue, Melbourne, Australia, July 12–14, 2018.*

32. J. Shin, P. Xu, A. Madotto and P. Fung, "Generating Empathetic Responses by Looking Ahead the User's Sentiment," in *2020 IEEE International Conference on Acoustics, Speech and Signal Processing, ICASSP 2020, Barcelona, Spain, May 4–8, 2020.*

33. R. W. Picard, "Affective Computing for HCI," in *Human-Computer Interaction: Ergonomics and User Interfaces, Proceedings of HCI International '99 (the 8th International Conference on Human-Computer Interaction), Munich, Germany, August 22–26, 1999, Volume 1, 1999.*

34. V. Scotti, R. Tedesco and L. Sbattella, "A Modular Data-Driven Architecture for Empathetic Conversational Agents," in *IEEE International Conference on Big Data and Smart Computing, BigComp 2021, Jeju Island, South Korea, January 17–20, 2021.*

35. A. Radford, J. Wu, R. Child, D. Luan, D. Amodei and I. Sutskever, "Language Models Are Unsupervised Multitask Learners," *OpenAI Blog*, vol. 1, p. 9, 2019.

36. D. P. Kingma, T. Salimans, R. Józefowicz, X. Chen, I. Sutskever and M. Welling, "Improving Variational Autoencoders with Inverse Autoregressive Flow," in *Advances in Neural Information Processing Systems 29: Annual Conference on Neural Information Processing Systems 2016, Barcelona, Spain, December 5–10, 2016.*

37. S. Park and J. Lee, "Finetuning Pretrained Transformers into Variational Autoencoders," *CoRR*, vol. abs/2108.02446, 2021.

38. E. Jang, S. Gu and B. Poole, "Categorical Reparameterization with Gumbel-Softmax," in *5th International Conference on Learning Representations, ICLR 2017, Toulon, France, April 24–26, 2017, Conference Track Proceedings, 2017.*

39. R. J. Williams, "Simple Statistical Gradient-Following Algorithms for Connectionist Reinforcement Learning," *Mach. Learn*, vol. 8, pp. 229–256, 1992.
40. I. Loshchilov and F. Hutter, "Decoupled Weight Decay Regularization," in *7th International Conference on Learning Representations, ICLR 2019, New Orleans, LA, USA, May 6–9, 2019*.
41. Y. Li, H. Su, X. Shen, W. Li, Z. Cao and S. Niu, "DailyDialog: A Manually Labelled Multi-turn Dialogue Dataset," in *Proceedings of the Eighth International Joint Conference on Natural Language Processing, IJCNLP 2017, Taipei, Taiwan, November 27–December 1, 2017, Volume 1: Long Papers*, 2017.
42. H. Rashkin, E. M. Smith, M. Li and Y.-L. Boureau, "Towards Empathetic Open-domain Conversation Models: A New Benchmark and Dataset," in *Proceedings of the 57th Conference of the Association for Computational Linguistics, ACL 2019, Florence, Italy, July 28–August 2, 2019, Volume 1: Long Papers*, 2019.
43. S. Zhang, E. Dinan, J. Urbanek, A. Szlam, D. Kiela and J. Weston, "Personalizing Dialogue Agents: I have a dog, do you have pets too?," in *Proceedings of the 56th Annual Meeting of the Association for Computational Linguistics, ACL 2018, Melbourne, Australia, July 15–20, 2018, Volume 1: Long Papers*, 2018.
44. E. Dinan, S. Roller, K. Shuster, A. Fan, M. Auli and J. Weston, "Wizard of Wikipedia: Knowledge-Powered Conversational Agents," in *7th International Conference on Learning Representations, ICLR 2019, New Orleans, LA, USA, May 6–9, 2019*.

III

Sentiment and Emotions

CHAPTER 4

EMOTRON

An Expressive Text-to-Speech

Cristian Regna, Licia Sbattella, Vincenzo Scotti, Alexander Sukhov, and Roberto Tedesco

DEIB, Politecnico di Milano, Milan, Italy

4.1 INTRODUCTION

Text-to-Speech (TTS) synthesis (or simply, speech synthesis) is the task of synthesizing a waveform uttering a given piece of text [1]. Like many other areas of the Artificial Intelligence (AI) field, Natural Language Processing (NLP) has been pervasively affected by Deep Learning (DL). The subsequent development of neural TTS systems (i.e., neural networks for speech synthesis) has dramatically advanced the state-of-the-art [2].

In the last few years, these neural network-based models evolved significantly, introducing neural vocoders to remarkably improve the quality of the synthesized speech [2]. They also introduced the possibility of conditioning the synthesized speech in different aspects, like the speaker's vocal timbre or prosodic style [2]. This latter aspect enables the TTS to be significantly expressive when uttering a sentence.

In this chapter, we introduce EMOTRON, a TTS system able to condition the synthesized speech on a given emotion[1]. The idea is to control the emotion expressed by the utterance by providing such emotion as an additional input to the neural network, during the synthesis process. During training, the network is updated to minimize speech synthesis and perceived emotion losses. In this way, we have the network learn to control the prosody (i.e., voice rhythm, stress, and intonation) necessary to deliver the emotional information through the uttered speech.

DOI: 10.1201/9781003296126-7

To assess the quality of our model, we conducted an evaluation based on human opinions. Listeners were asked to compare EMOTRON synthesized speech to real clips and to clips synthesized by a reference Tacotron 2 TTS we trained as a baseline. Compared to natural clips, those synthesized by EMOTRON and the baseline were always inferior. However, when compared to the baseline TTS, our results were slightly worse in terms of clarity of speech and clearly better in terms of perceived emotion.

We organize the rest of this chapter according to the following structure. In Section 4.2 we present the current approaches in the Deep Learning-based TTS development. In Section 4.3 we introduce our architecture for emotional TTS. In Section 4.4 we explain how EMOTRON is trained and used at inference. In Section 4.5 we present the corpora we employed to train our model. In Section 4.6 we describe the evaluation approach we followed to assess the quality of our model. In Section 4.7 we present and comment on the results of the evaluation of our TTS. In Section 4.8 we summarize our work and provide hints about possible future extensions.

4.2 RELATED WORKS

In this section we outline the main aspects concerning neural TTS synthesis models, and we provide details concerning the control of speech style in such systems.

4.2.1 Text-to-Speech Synthesis with Neural Networks

Deep Neural Networks enabled many new possibilities in developing TTS systems. In this chapter, we focus on approaches based on acoustic models [2]. TTS developed with this architecture are divided into two main components: acoustic model (also called spectrogram predictor) and vocode [1, 2]. The former component takes care of converting a sequence of graphemes[2] or phonemes[3] into a Mel-spectrogram; the latter component takes care of converting the Mel-spectrogram into a raw waveform, concluding the synthesis process.

A spectrogram is a visual way of representing a signal strength over time, at various frequencies. If such frequencies are passed through a mel filter (mathematical model trying to approximate the non-linear human sensitivity to various frequencies), the result is the Mel-spectrogram. Figure 4.1 shows an example. The horizontal axis shows time, the vertical axis shows mel frequencies, and color represents the strength of any given "point" (i.e., a given frequency bin at a given time instant).

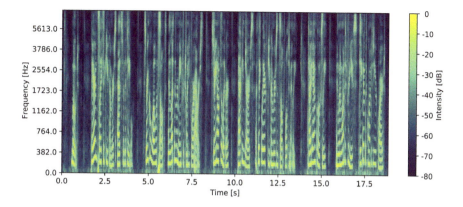

FIGURE 4.1   Visualization of a Mel-spectrogram.

Note that vowels, being almost harmonic, are represented as a set of "strips" (the biggest harmonic components, called formants), which both characterize each vowel and are unique for each speaker. Instead, unvoiced consonants, like "p" as in "pet," are shown like noise, appearing spread among all the frequency bins. Finally, voiced consonants, like "m" as in "man," are a mix of harmonics and noise.

The acoustic models for Mel-spectrogram prediction are commonly built using a Seq2Seq encoder–decoder architecture (e.g., Tacotron [3, 4], DeepVoice [5], FastSpeech [6]). The encoder projects the sequence of graphemes or phonemes into a sequence of hidden vectors. Instead, the decoder, either autoregressively or in parallel, generates the Mel-spectrogram. Usually, the alignment between encoded graphemes or phonemes and the Mel-spectrogram is done through an additional attention mechanism [7] working between the encoder and the decoder. Additionally, some of these architectures have a separate module to predict the stopping point of the generation process.

In particular, we are interested in the Tacotron 2 architecture [4], as this architecture has proven to be widely extensible [8] and re-usable [9]. Thus, we decided to enhance it with an emotion control module to make the generated voice more expressive.

To complete the speech synthesis pipeline, a vocoder is necessary. Neural vocoders have become a fundamental module of TTS models; they are necessary to synthesize a speech as clear as possible [1]. Compared to the previous approach, the Griffin-Lim algorithm [10], neural vocoders yield audio with fewer artefacts and higher quality. Available implementations are based on (dilated) convolutional neural networks (e.g., WaveNet [11],

WaveGlow [12], MelGAN [13, 14]) or recurrent neural networks (e.g., WaveRNN [15]). Moreover, they can be trained to work directly on raw waveforms [11] or on Mel-spectrograms [14]. For this work, we leveraged pre-trained implementations of WaveNet and WaveGlow (more on this in Section 4.4).

4.2.2 Controlled Speech Synthesis

Besides the obvious control on the synthesized speech given by the input text (what to say), there are multiple research lines focused on controlling further aspects of the synthesized speech through additional information. These additional aspects can be categorized into two groups: speaker (or timbre, i.e., who is speaking) and prosody (or style, i.e., how to speak) [2]. These two aspects are completely orthogonal and can be combined [8].

Speaker control allows disentangling the content from the speaker, making the overall TTS model more re-usable. The idea is to provide additional information on the speaker to the TTS. This approach allows leveraging multi-speaker datasets, while, previously, neural TTS used to be trained on single-speaker datasets [16]. To implement this kind of conditioning, it is sufficient to concatenate the hidden features extracted from a speaker recognition network to the hidden representation of the input text, as it was done for Tacotron 2 [17].

Prosodic control covers all the aspects concerning intonation, stress, and rhythm, which characterize how a sentence is uttered. These aspects also influence the emotion perceived by the listener. A reference step towards this kind of control on the speaking style is represented by the Global Style Token (GST) [16]. Instead of explicitly modelling the aspect characterizing the prosody, this model uses unsupervised style representations learnt alongside the synthesis model. In the original implementation, a Tacotron 2 model was extended with a separate encoder that extracted a vector representing the so-called style embedding; this vector is concatenated to the hidden representation of the input text to provide the decoder with the style information [16]. Notice that this kind of control works at a high level: the specific changes connected to a given style are learnt and implemented by the spectrogram generator.

In this work, we will focus on prosodic style control. As premised, we are interested in controlling low-level aspects concerning the style of speech by selecting at a high level the target emotion to express.

4.3 EMOTRON MODEL

In this section we depict the architectural details of the EMOTRON model for controlled speech synthesis. We describe the architecture of the spectrogram predictor and the architecture of the emotional capturer.

4.3.1 Text-to-Speech

We based the EMOTRON architecture on that of a Tacoron 2 [4]. The overall network is the same: an encoder–decoder architecture with location-sensitive attention. Similar to the GST variant [16], we introduced an additional linear transformation to manage the additional emotion input. The overall architecture is depicted in Figure 4.2; we re-used all the hyperparameters from the original implementation.

The input stream of characters $x = (x_1, \ldots, x_m)$ representing the text to utter is passed through the Character encoder: a stack of three convolutional layers and a final BiLSTM layer [18]. At this step, the output is a sequence of feature vectors.

We concatenated each hidden vector of the input character stream with the embedding of the target emotion. We extracted such embeddings through a linear transformation taking as input the categorical probability distribution of the desired emotion (predicted by the emotional capturer on a reference Mel-spectrogram; more on this in Section 4.3.2) $e = (e_1, \ldots, e_k)$,

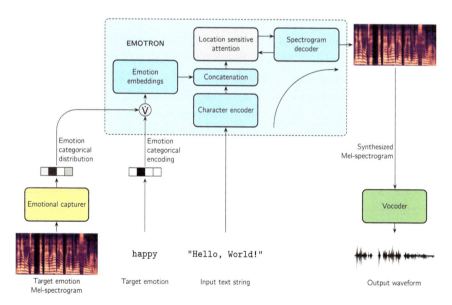

FIGURE 4.2   EMOTRON high-level architecture and data flow.

where $e \in [0, 1]^k \subseteq \mathbb{R}^k : \sum_{i=1}^k e_i = 1$. In this way, we can tell the network to imitate the emotion found in a reference audio clip. In particular, we considered $k = 4$ different emotions (namely, "neutral," "sadness," "anger," and "happiness") in our implementation, and generated 32-dimensional embeddings. Note that, alternatively, such distribution can actually be a one-hot encoding of the target emotion ($e \in \{0, 1\}^k$). In this way, we can tell the network to generate a specified emotion.

The hidden vector generated above is the input that guides the decoding of the output Mel-spectrogram. The alignment between the encoder and the decoder is realized through the location-sensitive attention [4]. All the hyper-parameters at this step were left unchanged.

The Spectrogram decoder is designed to work with a causal approach: it leverages all the Mel-spectrogram $Y_{t' < t}$ up to the current step t in the sequence to predict the next slice $\hat{y}_t$. The input Mel-spectrogram up to the latest generated step $Y_{t' < t}$ is passed through a stack of two pre-net layers, two LSTM [19] layers (aligned with the attention), and a final linear projection. The final output is further refined through a stack of five convolutional post-net layers. An additional linear projection to the latest LSTM output is used as a stop-net to signal the end of the generation process. All the hyper-parameters at this step were left unchanged.

4.3.2 Emotional Capturer

The emotional capturer plays two essential roles in the EMOTRON architecture (see Figure 4.3). On one hand, it takes care of extracting the perceived emotion probability distribution $e \in [0, 1]^k \subseteq \mathbb{R}^k : \sum_{i=1}^k e_i = 1$ from the Mel-spectrogram $Y \in \mathbb{R}^{C \times T}$ of a reference audio clip (where C is the number of frequency bins in the Mel-spectrogram and T is the length of the clip), so it is a discriminative neural network for emotion recognition. Thus, it allows replicating the perceived emotion from the reference clip during the synthesis process. On the other hand, this network's hidden features h_e can be leveraged to compute the style loss. As we explain in Section 4.4, this loss is fundamental to enforcing the emotion conditioning when training the whole network.

The emotional capturer network comprises a stack of six 2D convolutional layers that take the Mel-spectrogram as input. A Gated Recurrent Unit (GRU) layer and four fully connected layers complete the sequence of transformations. For further details about the architecture, refer to Figure 4.2.

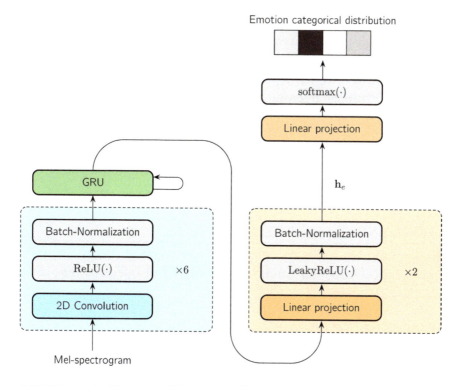

Emotion categorical distribution

FIGURE 4.3   Architecture of the emotional capturer.

The number of output channels in the convolutional layers doubles every two convolutional blocks (dashed blue box in Figure 4.3), starting from 32. Each 2D convolution uses a 3×3 kernel and a 2×2 stride. After every convolution, we apply ReLU(·) activation and batch normalization.

The GRU layer generates a sequence composed of 128-dimensional vectors (one for each time slice); we take only the last one to summarize the entire sequence.

Of the following two linear blocks (dashed orange box), the first one has, again, 128-dimensional vectors, while the last one yields 256-dimensional vectors. This feature vector represents $h_e = h_e(Y) \in \mathbb{R}^{256}$: the hidden representation of the whole sequence. Note that such linear blocks use a LeakyReLU(·) activation and dropout regularization.

Finally, a following liner projection yields the logits of the four considered emotions, and a softmax(·) activation allows to output their posterior probabilities e.

The overall architecture is agnostic of the actual labels it learns to discriminate. Conceptually, the emotion capturer acts as the Global Style

Token (GST) module from Tacotron 2 [16]. However, GST learns such labels through an unsupervised approach while training the entire model; the emotional capturer, instead, can be instructed on the target labels, independent of what they represent.

We also trained a second model using a semi-supervised learning approach, for the Tacotron 2 baseline model. We created alternative labels by clustering the clips in the dataset using common emotion discriminative features (more on this in Section 4.5). We used this pseudo-emotional capturer to add emotion control to the Tacotron 2 baseline model.

4.4 TRAINING AND INFERENCE

This section provides the details concerning the model's training (both EMOTRON and emotional capturer) and inference (i.e., audio synthesis).

4.4.1 Training Details

We designed the EMOTRON model to enforce emotion control on synthesized audio, as premised. We resorted to results coming from computer vision to achieve this goal [20]. We approached our problem following insights from image style transfer models: we trained our network to minimize, at the same time, a content loss $\mathcal{L}_{content}(\cdot)$ and a style loss $\mathcal{L}_{style}(\cdot)$, as described in the following equation; where x is the input sequence of characters, e is the target emotion categorical distribution, Y is the reference Mel-spectrogram, and $\hat{Y} = f_{TTS}(x, e \mid Y)$ is the output of EMOTRON, generated with guided decoding (more on this later).

$$\mathcal{L}_{TTS}(x, Y, e) = \mathcal{L}_{content}\left(Y, f_{TTS}(x, e \mid Y)\right) + \mathcal{L}_{style}\left(Y, f_{TTS}(x, e \mid Y)\right)$$

The content loss is the L_2 norm of the reconstruction error between target and predicted spectrograms, Y and $\hat{Y}$. The objective of this loss is to train the spectrogram predictor to output intelligible audio, and it is the usual loss used to train neural TTS models. We reported the formulation in the following equation, where C is the number of frequency bins forming the Mel-spectrograms (pre-defined and fixed), T is the number of time slices for the current reference Mel-spectrogram (different for each spectrogram we consider), while $y_{c,t}$ and $\hat{y}_{c,t}$ are the value of the c-th frequency bin of the t-th time slice, obtained from the reference Mel-spectrogram and generated by our model, respectively.

$$\mathcal{L}_{content}\left(\mathbf{Y},\hat{\mathbf{Y}}\right)=\left\|\mathbf{Y}-\hat{\mathbf{Y}}\right\|_2=\sqrt{\sum_{t=1}^{T}\sum_{c=1}^{C}\left(y_{c,t}-\hat{y}_{c,t}\right)^2}$$

At training time, the network uses guided decoding to generate $\hat{\mathbf{Y}}$. The Spectrogram decoder inside the EMOTRON model is designed to generate in an autoregressive manner, consuming as internal input the last generated Mel-spectrogram time slice. During training, however, we use the reference Mel-spectrogram time slice instead of recurring the output slice.

Instead, we used the style loss to enforce the emotion based (or emotion related, in the case of the pseudo-emotional capturer) style control on prosody. The idea behind this loss is to find space where it is possible to capture emotional information, like the hidden representation of a speech emotion discriminator. If the projection in such a space is a differentiable transformation (as in our case), it is possible to compute the style loss from the different hidden representations. We follow the approaches proposed for image style transfer [20, 21]; we compute the style loss as the L_2 norm of the difference between the Gram matrices $\mathbf{G}(\cdot) \in \mathbb{R}^{256 \times 256}$ obtained by the two hidden vectors of the reference and generated Mel-spectrograms. We reported the formulation of this loss in the following equation.

$$\mathcal{L}_{style}\left(\mathbf{Y},\hat{\mathbf{Y}}\right)=\left\|\mathbf{G}\left(h_e\left(\mathbf{Y}\right)\right)-\mathbf{G}\left(h_e\left(\hat{\mathbf{Y}}\right)\right)\right\|_2=\sqrt{\sum_{i=1}^{256}\sum_{j=1}^{256}\left(g_{i,j}-\hat{g}_{i,j}\right)^2}$$

We trained the EMOTRON network for almost 120 k update steps on mini-batches of 32 audio clips. We set the learning rate to $\eta = 10^{-3}$, with an exponential decay to 10^{-6}. To enforce regularization, we used weight decay with $\lambda = 10^{-6}$ and we clipped the maximum norm of the gradients to 1.

Such training procedure only affects the EMOTRON network $f_{TTS}(\cdot)$ (i.e., encoder–decoder) and the emotion embeddings.

Instead, the emotional (and pseudo-emotional) capturer $f_e(\cdot)$ was trained separately to discriminate among the considered emotions (or emotion-related clusters). We applied the usual negative log-likelihood loss computed through the cross-entropy on the target class. The following equation describes the loss, where e_i (with $i \in [1, k] \subseteq \mathbb{N}$, and $k = 4$ for the emotional capturer or $k = 3$ for the pseudo-emotional capturer), is the i th emotion associated with the speech signal represented through Y, while $f_e(\cdot)_i$ is the reconstructed probability for such ith emotion.

$$\mathcal{L}_e(e_i, Y) = -\ln P(e_i|Y) = -\ln f_e(Y_i)$$

4.4.2 Inference Details

During inference, the model leverages autoregressive decoding (i.e., it recurs the latest output time slice of the Mel-spectrogram as the next input of the decoder). Thus, instead of generating with a guided approach as in training and computing $\hat{y} = f_{TTS}(x, e\,|\,Y_{t'<t})$, it computes $\hat{y} = f_{TTS}(x, e\,|\,\hat{Y}_{t'<t})$ to generate the Mel-spectrogram $\hat{Y}$. The autoregressive process continues until the stop-net triggers the interruption (predicting a sufficiently high posterior "stop" probability).

Concerning emotion control, since we followed the same approach of GST, it is possible to obtain it in two ways. The former approach consists in providing a reference audio clip spectrogram to replicate its emotional style. The emotional capturer takes care of extracting the high-level information necessary to feed the model. This approach is followed during training and can be used for inference. Alternatively, the latter approach prescribes feeding the model with the categorical emotion and fetching the corresponding embedding to be concatenated to the textual features to condition the synthesis. We followed this approach only during inference.

Since EMOTRON outputs the Mel-spectrogram, to extract the raw audio waveform, generating the final synthesized speech, we leveraged a vocoder. During testing and inference, we leveraged the WaveNet vocoder [11] because of the higher audio quality (during training, to get samples faster, we employed WaveGlow [12]). Notice that we did not tie the model to a specific vocoder, and thus this final module can be freely substituted.

4.5 DATA

In this section, we present the corpus we employed to train the EMOTRON TTS model and the connected capturers. We organized the sections according to the target model.

4.5.1 Conditioned Speech Synthesis

The dataset we selected to train EMOTRON is the one released for the Blizzard Challenge 2013 [22]. It is a collection of audiobooks read by a single speaker, with high expressivity. The data set included more than 100 hours of recorded clips.

We selected this dataset for its size and high expressivity of uttered sentences. The high number of samples and their variety is crucial to have the network properly learning the different styles. Moreover, since we are not modelling multi-speaker properties, having a single speaker data set is crucial for convergence.

We retrieved two subsets of the original corpus: the "selected" version and the "full" version. The former is a selection of clips already pre-processed and paired with the transcription. The latter required some pre-processing to be usable, because the clips contained the reading of entire chapters. In particular, we did:

- Transcripts retrieval from the Project Gutenberg website[4].

- Transcripts alignment using the Aeneas forced aligner[5].

- Clips cropping to have easy-to-process small utterances.

- Post-processing to filter out clips shorter than 1 s and longer than 14 s.

The overall dataset resulted in 120 hours of recordings.

4.5.2 Emotion Recognition for EMOTRON

We required emotion labels on the speech synthesis dataset to train the emotional capturer. Since these labels were not available and due to the size of the dataset, we resorted to automatic systems. We leveraged two neural networks for emotion recognition from speech (and text): PATHOSnet [23] and CNN-MFCC [24]. We used these two systems for a more robust result.

PATHOSnet distinguishes among "neutral," "happiness," "sadness," and "anger"; CNN-MFCC can also identify "calm," "fear," "disgust," and "surprise." To have unified labels, we combined "calm" and "neutral," and "sad" and "fearful." Additionally, we removed "surprise" due to its low presence in the dataset.

We used these two networks to label the audio clips in the corpus by combining their predictions. Given the individual reported results, we applied the following rules to combine the predictions:

- If the predicted label is not "happy," use the prediction from PATHOSnet.

- Else use the prediction from CNN-MFCC.

4.5.3 Emotion-related Clusters Recognition for the Baseline Model

The original GST approach for expressive speech leveraged unsupervised learning to identify the style labels while training the speech synthesis model. Here instead, we propose splitting this step and learn pseudo-emotion labels.

Instead of relying on the hidden features learnt by the deep style model, we performed clustering on the audio clips and learnt to predict the cluster labels. We extracted features that correlate with emotion from the audio clips applied feature reduction and clustered clips on these representations.

We used the OpenSmile tool [25] to extract all 120 features including intensity, loudness, MFCCs, pitch and pitch envelope, probability of voicing, line spectral frequencies, and zero-crossing rate. We applied Principal Component Analysis (PCA) to reduce the features to 10 (retaining 90% of variance), and we used k-Means clustering with $k = 3$ (we also experimented with $k = 2$ and $k = 4$). Finally, we post-processed the result by deleting the clips too close to the boundaries between clusters.

4.6 EVALUATION APPROACH

This section outlines the approach we followed to assess the quality of the synthesized speech and the clarity of the emotion conditioning. We evaluated through a survey on a website we created specifically for this evaluation. The survey was composed of 12 questions, three for each considered emotion.

The website displayed a button to play the audio clip associated with the question and displayed the corresponding transcription of the clip. Under each clip, human listeners could provide their opinion scores about speech quality and recognized emotion. We aggregated the evaluations into Mean Opinion Scores (MOSs) we reported and commented on in Section 4.7.

To assess the quality of our model, we compared it against ground truth – real clips uttered by humans taken from the Blizzard Challenge 2013 corpus – and a baseline TTS model. As a baseline expressive TTS, we opted for Tacotron 2 with pseudo-emotion labels. This system uses the clusters extracted from the set of reference clips as target labels to add expressivity to the synthesized speech, conditioning the decoding on the identified style. We selected this model because it is based on Tacotron 2 and because it leverages a mechanism close to that of GST, which is considered state-of-the-art in terms of expressive TTS (the difference is that our approach applies the unsupervised step of clustering separately from the speech synthesis training). Additionally, we considered it also because it leverages the same core architecture of EMOTRON, enabling a direct comparison between the pseudo-emotion clusters and actual emotion labels used for conditioning.

To achieve the best results from EMOTRON, we leveraged the WaveNet vocoder to convert the Mel-spectrogram into a waveform. For better comparison, we used the same vocoder for the baseline TTS with pseudo-labels.

4.6.1 Speech Quality

To assess the quality of the EMOTRON spectrogram predictor, we used the human listener's MOS. We asked each listener to rate the quality of the audio clip on a 1-to-5 scale with a unit increment. The idea is to indirectly evaluate the quality of the spectrogram predictor by evaluating the audio quality.

To help listeners in the evaluation, we provided the following explanation for the scoring system:

1. **Bad**, unrecognizable speech.

2. **Poor**, speech is barely recognizable.

3. **Fair**, acceptable quality of speech, small errors allowed.

4. **Good**, speech with proper pronunciation and quality.

5. **Excellent**, perfect speech, sounds natural and expressive.

4.6.2 Emotional Clarity

To assess the quality of the EMOTRON emotion conditioning modules and the overall emotional clarity of the synthesized speech, we calculated the human listener's average emotion recognition accuracy. We asked the listeners to associate one of the four possible emotion labels to each clip: neutral, sad, angry, and happy.

We synthesized EMOTRON's clips conditioning it on the target emotion expressed in a reference clip from the ground truth. The baseline TTS was conditioned on the emotional labels detected by the pseudo-emotional capturer. We used the same clips of the speech quality evaluation, including the ground truth.

4.7 RESULTS

We collected responses from 54 listeners during the three days the survey website was online. In the following, we report and comment on the final scores.

TABLE 4.1   Results of Human Evaluation of Speech Quality

	Speech Quality: MOS				
	Neutral	Sadness	Anger	Happiness	Average Score
Ground truth	4.50	4.75	4.20	4.60	4.53
Tacotron 2 w/ pseudo-emo.	4.12	3.89	4.08	3.77	3.96
EMOTRON	4.10	4.00	3.71	3.65	3.87

4.7.1 Speech Quality

We reported the MOS collected during the evaluation in Table 4.1. We reported the quality scores averaged emotion-wise and averaged on the entire support.

Ground truth consistently outperforms both TTS models. This result is not surprising considering that these clips are actual human voice samples. Ground truth relative performances are 14.4 and 17.1% better than Tacotron 2 with GST and EMOTRON, respectively. Looking at the emotion-wise breakdown, we can see that the highest score was on "sadness" clips and the worst score on "anger" clips; yet, any single ground truth score is higher than any of the two TTS models.

EMOTRON underperforms the baseline TTS in the audio quality evaluation. However, the relative performance difference with respect to this baseline TTS is only –2.3%, meaning that the two systems deliver audio clips with very similar quality. Moreover, the spectrogram predictor is the same between the two models. This is probably due to the fact that the style conditioning system we developed affects audio quality independent of the target style.

4.7.2 Emotional Clarity

We reported the accuracy scores collected during the evaluation in Table 4.2. We also reported the clarity scores averaged emotion-wise and averaged on the entire support.

TABLE 4.2   Results of Automatic Emotion Evaluation of Emotional Clarity

	Emotion Clarity: Recognition Accuracy (%)				
	Neutral	Sadness	Anger	Happiness	Average Score
Ground truth	89	78	63	80	78
Tacotron 2 w/ pseudo-emo.	54	48	49	41	48
EMOTRON	70	51	55	46	56

We leverage the listener's emotion recognition accuracy as a scoring function, as premised. If we consider the results on ground truth, accuracy scores are similar to those of other works, where humans reported a recognition accuracy of ~70.0% [26].

Similar to speech quality, ground truth outperforms both TTS models in emotional clarity. Ground truth relative performances are 62.5 and 39.3% better than the baseline TTS and EMOTRON, respectively. Looking at the emotion-wise breakdown, we can see that the highest score was on "neutral" clips for all systems. Instead, the worst score is on "anger" clips for ground truth, while both TTS models perform worse on "happiness" clips.

Unlike speech quality, EMOTRON outperforms the baseline TTS in the audio quality evaluation, and it does so by a consistent margin. The relative performance difference with respect to this baseline TTS is 16.7%. Since the two TTS models share the same spectrogram predictor architecture and were trained on similar datasets, we can hypothesize that the unsupervised model learnt by clustering the clips did not reflect the division among different emotions despite the selected features. To fully understand the reason behind this difference, an ablation study on the two networks would be necessary.

4.8 CONCLUSION

In this chapter, we presented EMOTRON: a TTS with emotion conditioning. EMOTRON builds on top of a well-known neural TTS architecture (Tacotron 2) to synthesize expressive speech. We extended the base architecture to enhance the expressivity of the uttered text by training it to synthesize audio given an input text and the emotion to display.

During training, we used a combination of Mel-spectrogram reconstruction and style losses; in this way, we enforced generative capabilities and emotion control on EMOTRON. We used the Mel-spectrogram reconstruction loss to have the model learn generative capabilities, together with the style loss to measure the distance between the enforced emotion and the desired one. We computed it through a separate neural network designed to discriminate among emotions from speech. We used the style loss to have the model learn how to add expressivity in the synthesized speech.

To assess the quality of our model, we resorted to human evaluation to measure speech quality and emotional clarity. In the evaluation, we compared EMOTRON with natural human speech and another TTS for expressive speech built on the same architecture leveraged by EMOTRON. The baseline system uses pseudo-emotional labels learnt through clustering.

Both TTS models performed worse than human speech in terms of synthesized audio quality and emotional clarity, thus leaving space for many improvements. Instead, the two TTS performed comparably in speech quality (despite the baseline TTS being slightly better). EMOTRON, however, outperformed the baseline TTS by a consistent margin in terms of emotional clarity. This result highlighted that the use of unsupervised labels (as we did for the pseudo-emotional capturer), despite being useful in general, provided worse results than an approach based on supervised learning (like the one leveraged by EMOTRON) when such labels represent emotions.

NOTES

1. The implementation is available at: https://github.com/Sashorg/Emotional_TTS-master
2. A grapheme is "the smallest meaningful contrastive unit in a writing system." In other words, it is a written symbol that represents a sound; it could be represented by a single letter or a sequence of letters, such "sh."
3. A phoneme is any of the perceptually distinct units of sound in a specified language that distinguish one word from another, for example "p," "b," "d," and "t" in the English words "pad," "pat," "bad," and "bat."
4. http://gutenberg.org/
5. https://www.readbeyond.it/aeneas/

REFERENCES

1. D. Jurafsky and J. H. Martin, "Speech and language processing: an introduction to natural language processing, computational linguistics, and speech recognition (3rd Edition)," https://web.stanford.edu/~jurafsky/slp3/, 2022.
2. X. Tan, T. Qin, F. K. Soong and T.-Y. Liu, "A Survey on Neural Speech Synthesis," *CoRR*, vol. abs/2106.15561, 2021.
3. Y. Wang, R. J. Skerry-Ryan, D. Stanton, Y. Wu, R. J. Weiss, N. Jaitly, Z. Yang, Y. Xiao, Z. Chen, S. Bengio, Q. V. Le, Y. Agiomyrgiannakis, R. Clark and R. A. Saurous, "Tacotron: Towards End-to-End Speech Synthesis," in *Interspeech 2017, 18th Annual Conference of the International Speech Communication Association, Stockholm, Sweden, August 20–24, 2017.*
4. J. Shen, R. Pang, R. J. Weiss, M. Schuster, N. Jaitly, Z. Yang, Z. Chen, Y. Zhang, Y. Wang, R. J.-S. Ryan, R. A. Saurous, Y. Agiomyrgiannakis and Y. Wu, "Natural TTS Synthesis by Conditioning Wavenet on MEL Spectrogram Predictions," in *2018 IEEE International Conference on Acoustics, Speech and Signal Processing, ICASSP 2018, Calgary, AB, Canada, April 15–20, 2018.*

5. W. Ping, K. Peng, A. Gibiansky, S. Arik, A. Kannan, S. Narang, J. Raiman and J. Miller, "Deep Voice 3: Scaling Text-to-Speech with Convolutional Sequence Learning," in *6th International Conference on Learning Representations, ICLR 2018, Vancouver, BC, Canada, April 30–May 3, 2018, Conference Track Proceedings*, 2018.

6. Y. Ren, Y. Ruan, X. Tan, T. Qin, S. Zhao, Z. Zhao and T.-Y. Liu, "FastSpeech: Fast, Robust and Controllable Text to Speech," in *Advances in Neural Information Processing Systems 32: Annual Conference on Neural Information Processing Systems 2019, NeurIPS 2019, December 8–14, 2019, Vancouver, BC, Canada*, 2019.

7. D. Bahdanau, K. Cho and Y. Bengio, "Neural Machine Translation by Jointly Learning to Align and Translate," in *3rd International Conference on Learning Representations, ICLR 2015, San Diego, CA, USA, May 7–9, 2015, Conference Track Proceedings*, 2015.

8. R. J. Skerry-Ryan, E. Battenberg, Y. Xiao, Y. Wang, D. Stanton, J. Shor, R. J. Weiss, R. Clark and R. A. Saurous, "Towards End-to-End Prosody Transfer for Expressive Speech Synthesis with Tacotron," in *Proceedings of the 35th International Conference on Machine Learning, ICML 2018, Stockholmsmässan, Stockholm, Sweden, July 10–15, 2018*.

9. A. Favaro, L. Sbattella, R. Tedesco and V. Scotti, "ITAcotron 2: Transfering English Speech Synthesis Architectures and Speech Features to Italian," in *4th International Conference on Natural Language and Speech Processing, Trento, Italy, November 12–13, 2021*.

10. D. W. Griffin and J. S. Lim, "Signal estimation from modified short-time Fourier transform," in *IEEE International Conference on Acoustics, Speech, and Signal Processing, ICASSP '83, Boston, Massachusetts, USA, April 14–16, 1983*, 1983.

11. A. van den Oord, S. Dieleman, H. Zen, K. Simonyan, O. Vinyals, A. Graves, N. Kalchbrenner, A. W. Senior and K. Kavukcuoglu, "WaveNet: A Generative Model for Raw Audio," in *The 9th ISCA Speech Synthesis Workshop, Sunnyvale, CA, USA, 13–15 September 2016*.

12. R. Prenger, R. Valle e and B. Catanzaro, "Waveglow: A Flow-based Generative Network for Speech Synthesis," in *IEEE International Conference on Acoustics, Speech and Signal Processing, ICASSP 2019, Brighton, United Kingdom, May 12–17, 2019*.

13. K. Kumar, R. Kumar, T. de Boissiere, L. Gestin, W. Z. Teoh, J. Sotelo, A. de Brébisson, Y. Bengio and A. C. Courville, "MelGAN: Generative Adversarial Networks for Conditional Waveform Synthesis," in *Advances in Neural Information Processing Systems 32: Annual Conference on Neural Information Processing Systems 2019, NeurIPS 2019, December 8–14, 2019, Vancouver, BC, Canada*, 2019.

14. G. Yang, S. Yang, K. Liu, P. Fang, W. Chen and L. Xie, "Multi-Band Melgan: Faster Waveform Generation For High-Quality Text-To-Speech," in *IEEE Spoken Language Technology Workshop, SLT 2021, Shenzhen, China, January 19–22, 2021*.

15. N. Kalchbrenner, E. Elsen, K. Simonyan, S. Noury, N. Casagrande, E. Lockhart, F. Stimberg, A. van den Oord, S. Dieleman and K. Kavukcuoglu, "Efficient Neural Audio Synthesis," in *Proceedings of the 35th International Conference on Machine Learning, ICML 2018, Stockholmsmässan, Stockholm, Sweden, July 10–15, 2018*.

16. Y. Wang, D. Stanton, Y. Zhang, R. J. Skerry-Ryan, E. Battenberg, J. Shor, Y. Xiao, Y. Jia, F. Ren and R. A. Saurous, "Style Tokens: Unsupervised Style Modeling, Control and Transfer in End-to-End Speech Synthesis," in *Proceedings of the 35th International Conference on Machine Learning, ICML 2018, Stockholmsmässan, Stockholm, Sweden, July 10–15, 2018*.

17. Y. Jia, Y. Zhang, R. J. Weiss, Q. Wang, J. Shen, F. Ren, Z. Chen, P. Nguyen, R. Pang, I. Lopez-Moreno and Y. Wu, "Transfer Learning from Speaker Verification to Multispeaker Text-To-Speech Synthesis," in *Advances in Neural Information Processing Systems 31: Annual Conference on Neural Information Processing Systems 2018, NeurIPS 2018, December 3–8, 2018, Montréal, Canada*, 2018.

18. M. Schuster and K. K. Paliwal, "Bidirectional Recurrent Neural Networks," *IEEE Trans. Signal Process*, vol. 45, pp. 2673–2681, 1997.

19. S. Hochreiter and J. Schmidhuber, "Long Short-Term Memory," *Neural Comput*, vol. 9, pp. 1735–1780, 1997.

20. L. A. Gatys, A. S. Ecker and M. Bethge, "Image Style Transfer Using Convolutional Neural Networks," in *2016 IEEE Conference on Computer Vision and Pattern Recognition, CVPR 2016, Las Vegas, NV, USA, June 27–30, 2016*.

21. J. Johnson, A. Alahi and L. Fei-Fei, "Perceptual Losses for Real-Time Style Transfer and Super-Resolution," in *Computer Vision – ECCV 2016 – 14th European Conference, Amsterdam, The Netherlands, October 11–14, 2016, Proceedings, Part II*, 2016.

22. S. King and V. Karaiskos, "The Blizzard Challenge 2013," *festvox*, 2013.

23. V. Scotti, F. Galati, L. Sbattella and R. Tedesco, "Combining Deep and Unsupervised Features for Multilingual Speech Emotion Recognition," in *Pattern Recognition. ICPR International Workshops and Challenges – Virtual Event, January 10–15, 2021, Proceedings, Part II*, 2020.

24. M. G. de Pinto, M. Polignano, P. Lops and G. Semeraro, "Emotions Understanding Model from Spoken Language Using Deep Neural Networks and Mel-Frequency Cepstral Coefficients," in *2020 IEEE Conference on Evolving and Adaptive Intelligent Systems, EAIS 2020, Bari, Italy, May 27–29, 2020*.

25. F. Eyben, F. Weninger, F. Gross and B. Schuller, "Recent Developments in OpenSMILE, the Munich Open-Source Multimedia Feature Extractor," in *Proceedings of the 21st ACM International Conference on Multimedia*, New York, NY, USA, 2013.

26. V. Chernykh, G. Sterling and P. Prihodko, "Emotion Recognition from Speech with Recurrent Neural Networks," *CoRR*, vol. abs/1701.08071, 2017.

IV

Fake News and Satire

CHAPTER 5

Distinguishing Satirical and Fake News

Anna Giovannacci and Mark J. Carman

Politecnico di Milano, Milan, Italy

5.1 INTRODUCTION

In recent years, the spread of fake news has become a worldwide phenomenon, bringing instability in both social and political situations. The phenomenon is hard to control, and its origin can be traced back to the birth and expansion of digital social networks. The global internet provides for exceedingly fast dissemination of fake news, which resembles a tidal wave in its ability to reach and infiltrate any space. For even the most highly educated individuals, it has become a challenge to assess the veracity of the articles that we read daily on our screens.

Humans like to laugh. The cause of humor can be varied: from images and videos representing funny situations to sentences containing jokes and puns that are humorous by virtue of their wordplay or meaning. For a long time, researchers have tried to provide definitions and understand what generates humorous content and funny emotions. In this chapter we won't, however, try to model satire or sarcasm with rules or definitions; instead we will let a statistical model learn from a large number of written examples what is funny and what is not.

To be more precise, the model developed in this chapter will be trained to tell apart what is real and what has been fabricated with the intent to make the reader laugh, or to manipulate the perception or emotion of the reader – in other words, to distinguish between *real*, *satirical*, and *fake* news. In addition to this classification task, we

DOI: 10.1201/9781003296126-9

perform a novel investigation in the use of techniques from eXplainable Artificial Intelligence (xAI) for discovering word patterns and rhetorical figures that can be linked to the different types of content: fake, real, or sarcastical texts.

5.1.1 Origins of Satire in Political Discourse

Given that this work concerns satire, it is useful to discuss when the genre was born and how its characteristics have evolved over time. Many literary genres, such as tragedy and comedy, are considered to have originated in ancient Greece, and this is also the case for satire, with the comic Aristophanes sometimes considered the progenitor of the genre. The origin of the word "satire" is quite different, with the term originating in ancient Rome; as the author Quintilian stated in the Institutio Oratoria: *satura quidem tota nostra est* (satire is really totally ours). There are many theories about how the word came into use. The most accredited one associates the term with the expression *Satura lanx*, which in Latin indicates the mixed dish of first fruits of the earth destined for the gods. This leads the reader to identify satire as a sort of *miscellanea*, a text that was treating a lot of different arguments [1]. Roman satire is one of the first examples of what today we would call controversy. The objectives were precise and targeted, with disparate motivations, and there were not the same limitations as other genres: writers were free to follow their own style from beginning to end. Looking to Italian literature, satire first followed the model set by the Roman authors, but subsequent to the capture of Constantinople by Mohammed II, Hellenic-style intellectuals took refuge in the West and the Latin satirical genre merged with the Greek satirical drama. Satire became in a way more violent and aggressive than it was before, and this newborn mixture became the satire that we love and practice today.

Satire has always targeted specific people, due to their particular behavior (e.g., abusive) or lifestyle (e.g., non-conforming to social norms). Over time, satire shifted its focus to politics, and in general, targeted people in authority. It became the voice of the people, an irreverent and irritating voice which highlights the contradictions of power and delineates its defects. Contemporary examples of political satire can be found on the Web in the form of satirical news sites, such as The Onion. Under the Italian jurisdiction, the Court of Cassation gave its own legal definition of satire in 2006, which seems quite appropriate and self-explanatory, so we report its translation to English here:

> *It is that manifestation of thought at times of the highest level that over time has taken on the task of "**castigare ridendo mores**," or of indicating to the public opinion criticizable or execrable aspects of people, in order to obtain, through the aroused laughter, a final result of ethical, corrective that is towards good.*
>
> —*Prima sezione penale della Corte di Cassazione, sentenza n. 9246/2006*

This citation is remarkable in how concisely it explains satire with the Latin phrase *castigare ridendo mores*, which means "punish customs with laughter."

5.1.2 Satire vs. Fake News

In the past, satire was widespread in the media that was available at the time: first newspapers and then television. In more recent times, with the spread of the internet, satire has also extended to this medium. Similar to what happens with all content on the internet, even satirical texts spread faster now than they did before. Moreover, the sources are no longer only newspapers, and news are no more interpreted only by journalists and experts, but also by people from all walks of life, occupations, and educational levels [2]. Indeed, satire has become more pervasive in everyday life and as a consequence likely influences more the lives of people than in the past.

At the same time, it is easy to understand how fake news has become a worrying and dangerous phenomenon. The speed of its diffusion has greatly increased, and therefore also its potential for damage [2]: From convincing people to seek incorrect or harmful medical treatment for their illness to spreading untruths about individuals that affect their private or political life.

Thus, in this chapter, we decided to focus on detecting and distinguishing between satirical, fake, and real news. There are many cases where distinguishing between these types of content is critical. For example, during times of emergency, like the pandemic, or climate crises, like a flood or earthquake, it could be important to see how satirical, fake, and real news subjects spread over the internet and affect the population. Alternatively, certain individuals (e.g., politicians) or companies (especially retail ones) may need to understand how they are perceived online, which will depend on whether comments mentioning them are real, fake, or satirical.

5.2 RELATED WORK

We now discuss state-of-the-art text classification methods used for fake news and satire detection. We introduce and explain the model that we trained, followed by the methods that we used to obtain explanations for the predictions.

As noted in [3], Natural Language Processing (NLP) techniques have evolved from *Symbolic NLP*, which used formal languages and set of rules together with dictionary lookups to classify text (struggling with the many ambiguities of natural language), to *Statistical NLP*, which used Machine Learning (ML) techniques and big corpora of text samples to overcome ambiguities, to *Deep Learning NLP*, which combines deep Neural Networks with massive quantities of data to immensely improve language prediction tasks. We follow the Deep Learning (DL) approach, making use of the standard BERT [4] models.

5.2.1 Fake News and Satire Detection

We distinguish among three types of content:

- *Real news articles*: text containing facts that can be verified to be true.

- *Satirical news articles*: text containing features such as sarcasm or irony in order to ridicule someone or something. In general, satirical articles cause no harm, except perhaps if they are misunderstood as true.

- *Fake news articles*: text in which false information is deliberately presented as though it were true. Such misinformation can cause many problems, such as misleading voters to influence elections or providing incorrect medical advice leading to harmful treatments.

With regard to satire detection, most previous works have tried to distinguish between real and sarcastic news. The techniques needed to cope with the fact that if the samples are divided by their sources, and each source provides only one type of content (either satirical or real), then the model tends to learn the style of the source, rather than generic indicators of satire. For example, in Rogoz et al. [5], the sources were split into training and test sets, in order to verify the accuracy of the classifier without letting it simply overfit to a single source. Another way to overcome this issue was proposed in [6], and it consisted in building an adversarial component to detect the source. Some features, however, were overlapping between satire and source classification, suggesting that the two tasks are not completely independent.

As for almost all tasks that are currently tackled with DL, also fake news detection was and has been approached with statistical methods and Machine Learning (ML). According to a recent survey [7], there are four common approaches for distinguishing between fake and real news: *style-based* techniques consider characteristics defining the style of the article, while *source-based* techniques assign credibility to each different source, *knowledge-based* approaches rate content by extracting claims and checking whether they are consistent with known facts, and finally *propagation-based* approaches measure characteristics of the diffusion of the information on the social network.

5.2.2 Explainable AI Techniques

Artificial Intelligence (AI) is a field of study which aims to augment machines with intelligence, itself a rather complex and not necessarily well-defined concept. ML aims to let machines learn from past experience and adapt to their environment. ML algorithms that make predictions based on evidence can be seen as *white boxes* (if you can look inside them) or *black boxes* (if you only see the prediction). The purpose of XAI is to make prediction algorithms understandable by humans, by providing an explanation for each prediction that allows the user to understand why the model predicted as it did.

The reasons for providing explanations are many. For applications of AI in healthcare, a practitioner may need an explanation for a disease prediction before they can believe the diagnosis, while a patient may need the explanation in order to comprehend a prognosis (or to change their behavior appropriately to reduce their risk). Similarly, in biology, the explanation of why an algorithm predicts a particular genetic condition might help a researcher in determining what genetic features are "responsible" for the result.

For DL models, there exist a variety of different explanation methods depending on the prediction model used. For example, for Convolutional Neural Networks applied to images, Class Activation Maps (CAMs) are often used in order to show which part of the image contributed most to the prediction [8]. An ideal explanation has three main features as noted in Ribeiro et al. [9]; it should be *interpretable* by the user, exhibit *local fidelity* explaining how the model behaves around the analyzed instance, and *model-agnostic* so as to produce explanations for any model.

In this work, we use a *gradient-based* approach to produce saliency maps over text, as described in Nielsen et al. [10]. Before going deeper into these methods, we note that for neural networks, and specifically text classification, *perturbation-based* methods also exist which modify the inputs and observe how the output changes, such as Lime [9]. The main problem with such techniques is the computational footprint compared to gradient-based methods. There exist a number of different gradient-based methods [11], with the general idea behind them is seeing how the changes in input affect the output. The one utilized for this chapter is an extension of the *Saliency* [12] method, called *InputXGradient* [13]. This method multiplies the gradient of the input by the size of the input embedding.

5.2.3 Questions Addressed in This Study

We focus on identification of sarcasm in shorter texts and on the use of multilingual models (since past approaches for this task were almost invariably monolingual). Building on this, we ask the following research questions:

1. Is it possible, using modern tools, to correctly classify short sentences as satirical vs. non-satirical?

2. Is it possible to build a classifier that works well across languages and cultures?

3. Is the performance of a monolingual classifier always better than the multilingual one?

   Furthermore, it seemed interesting and valuable to apply explainability methods to the trained neural models, in order to observe and comment on which features had the biggest influence on the predictions for our test samples. In particular, we decided to apply a gradient-based explainability method with our discriminative BERT models, even though they are generally applied with generative models, like GPT-2 [14], resulting in another research question:

4. Making use of gradient-based methods, is it possible to produce an explanation that is meaningful for the context and therefore build a tool that is human-interpretable and useful?

While investigating this, it was natural to look at the difference between sarcastic news and fake news. Fake news spreads on the internet due to the increasing number of platforms to post content on and the astonishing social connections between users, but they present some similarities in style and content with sarcastic articles or tweets, although the former are perhaps less popular. Hence, the final research questions that we added were:

5. Is it possible, using modern tools, to correctly classify a text as fake, real, or satirical news?

6. Is it possible to display clearly how the prediction was made by the model and which parts of the sentences were responsible for it, in order to identify and see patterns in words and phrases and to analyze what linguistic features cause the model to make correct predictions or to miss satire (i.e., make a mistake)?

5.3 METHODS

Following on the discussion above regarding state-of-the-art techniques for modelling textual data, we focus on the use of Bidirectional Encoder Representations from Transformers (BERTs). However, with respect to many research projects making use of BERT for similar tasks, the elements of novelty in this work can be summarized as follows:

- **Highly multilingual**: The number of languages analyzed in this work is elevated and not common in other works. In fact, we fit the model using tweets in English, Italian, French, German, and Spanish.

- **Distinction between real, satirical, and fake news**: In most of the previous works, the main classification tasks undertaken were Satire vs. Fake or Fake vs. Real, so that the outcome was binary.

- **Explainability with gradient methods**: usually other methods for explanations are used in BERT, like non-negative matrix factorization. Here, we have used gradient-based methods, and the saliency maps that we build will be further used to identify textual patterns and do some error analysis.

In the following sections, we briefly introduce the approach taken and the reasoning for the research questions posed.

5.3.1 Multilingual Satirical Text Classification with Transformers

In order to tackle the first three research questions regarding multilingual classification of satirical vs. not-satirical text, we adopted a multilingual BERT classifier, implemented in TensorFlow [15]. From the beginning, the intention was to build a multilingual dataset, which we collected from different sources reported in this section. One issue that arises from using a multisource dataset is the fact that the model could learn the features of the sources rather than the pattern of the irony, an aspect that will be discussed further later. The data gathering part was carried out after careful consideration of this problem. The intention was to try and build a multidomain dataset from several sources, while using as a test set a sample of tweets taken from sources that were not used in the training phase, as done by [5].

5.3.2 Three-Way Classification: Satirical vs. Fake vs. Real

The next step of the work was including a different label for the model to be classified: concentrating on the English language, we prepared a new dataset in order to distinguish between fake, real and satirical news, trying as much as we could to differentiate sources and to enlarge the knowledge base of the model (data sources: see Tables 5.1 and 5.2). This task was approached both as regards long texts and as regards short texts, and we focused on the English language, due to the greater availability of data.

TABLE 5.1 Twitter Sources Used for Training, Divided by Language

Language	Real News Sources	Satirical News Sources
Italian	ANSA, *Corriere della Sera*	*Spinoza, Lercio*
English	*The New York Times*, CNN	The Onion, NewsThump
French	*Le Figaro*	*Charlie Hebdo*
Spanish	EFE	*El Mundo Today*
German	DPA	*Nebelspalter*

TABLE 5.2 Twitter Sources Used for Testing, Divided by Language

Language	Real News Sources	Satirical News Sources
Italian	*la Repubblica*	Kotiomkin
English	HuffPost	The Daily Mash
French	*Le Courrier*	*Le Canard Enchaîné*
Spanish	EL PAÍS	*El Jueves*
German	SPIEGEL Eil	*Titanic*

5.3.3 Gradient-Based Explanations

The next research question asked whether by using a gradient-based method the learning process of the model could be explained and better understood by humans. We took inspiration from [16] to build a saliency map over input words that could show which were the most important words for each prediction. The theoretical concepts in this approach have been explained in Section 2.2 and here they will be depicted in simpler words. The steps of the process are:

- Take an input and obtain its embedding, by looking up its input IDs in the embedding matrix.

- With respect to this input embedding, calculate the gradient of the loss for the prediction of the classifier.

- A vector with the size of the input sequence is obtained, and every element has a different weight according to how much the word in the initial sequence has importance for the prediction.

This kind of explanation method has primarily been used for explaining predictions from a GPT-2 model, but in this work, we use it to explain predictions from a BERT model.

5.3.4 Interactive User Interface-Based Analysis

To make the explanations available, a web application was built using Flask. The intention was to let users insert the text that they want to be classified, letting them choose the scope they prefer, and then give them back the prediction, if possible, the real label and the predicted one, and, moreover, the explanation about the prediction.

5.4 EXPERIMENTS

In this section, we detail the experiments performed to compare the performance of various classification models. We start with a binary short-text classifier, comparing a multilingual classifier with a monolingual one. After that, we consider the three-label classifier, still for short text, and finally we move on to assess the performance of the long text classifier for articles, also with three labels.

5.4.1 Short-Text Classification

As explained above, the first aim of this work was to distinguish between satirical and non-satirical short texts. We investigate Twitter data in

TABLE 5.3   Performance of Monolingual vs. Multilingual Tweet Classifier

Model	Precision	Recall	Accuracy
Multilingual	0.77	0.80	0.79
Monolingual	0.88	0.80	0.85

particular but believe that the results should generalize to other types of short social media posts. We investigated whether a large multilingual dataset would provide the same level of performance as a smaller monolingual dataset, but as can be seen in the Table 5.3, precision was higher for the monolingual classifier.

It seemed natural that a satire classifier should also be able to detect fake or tendentious tweets (i.e., those promoting an incorrect or controversial viewpoint). At first, we investigated adding an English fake news dataset to the multilingual tweet dataset, but the accuracy was low on the fake samples and the features that the model learned were biased towards the samples. So, we decided to switch to an English-only fake news dataset, obtaining better accuracy and better separation between these samples. Performances are shown in Figure 5.1.

Looking at the confusion matrix in Figure 5.1, we see that fake and satirical news are more often confused with each other than with the real news. Accuracy was reasonable at around ~82%. As a test set for fake headlines, we used tweets from the dataset available in [17]. The analysis of the results will be explained in the remainder of this chapter.

5.4.2 Long-Text Classification

Classifiers trained on shorter texts are unlikely to generalize to longer texts, so we then investigated training and evaluating a classifier for longer

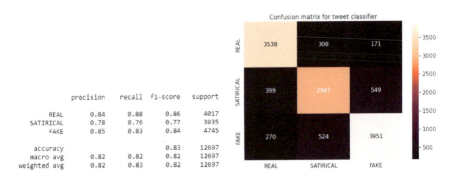

FIGURE 5.1   Validation performance (left) and confusion matrix (right) for short text classification of tweets.

article-length texts. The task was to classify long texts into the usual three labels (real, fake, satirical). The model was trained for 20 epochs, with a learning rate of 10^{-3} and a batch size of 8. With respect to the previous models, more training time was necessary to obtain the same level of accuracy. This could be due to the article source being less significant, or because the texts are longer. The dataset was obtained by the following sources:

- Fake news came from a challenge on Kaggle[1] and regarded the American elections of 2016[2].

- Real news was also taken from Kaggle[3] and consisted of articles listed on the website *AllSides.com/unbiased-balanced-news*, which were articles from different news sources over the time period: 31/05/2017 to 19/02/2018.

Looking at the confusion matrix for long texts in Figure 5.2, one can see that satirical samples were mostly confused with real samples, while almost none of them were confused with fake ones. Fake samples, meanwhile, were confused almost at the same rate with real and sarcastic ones. This suggests directions for future work, that will be discussed further later.

5.4.3 Analyzing Common Failure Cases for Detecting Satire

After training and evaluating the models, the next step was to try to understand *why* the predictions had that outcome and what are the main factors that the classifier considers. This task comprehends why the model makes a certain kind of mistakes, how overfitting (if present) is influencing the results, and to what extent satirical news and fake news are similar and

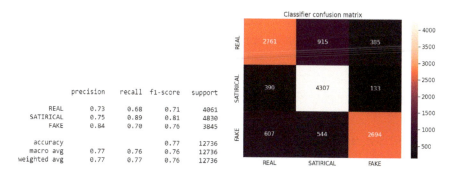

FIGURE 5.2   Validation performance (left) and confusion matrix (right) for long text classification.

why it is so. As discussed previously, a gradient-based method for explaining the predictions was adopted.

5.4.3.1 Irony, Temporal Aspects, and Subtlety

In this section, we analyze the most common errors that the classifier makes and try to explain its behavior. We start with the short-text classifier and then we move on to the long-text classifier.

For the short-text classifier, to compute reliable metrics for the evaluation of the performance, we made use of a different dataset (from the training set) called TWITTIRÒ [18], which was made up of Italian tweets from three different sources. Given these input data (which are around 1400 samples) we assess the performance of the model on these new tweets. These tests were performed using the model trained on the multilingual dataset. Some interesting points were found:

1. The majority of the tweets in the dataset were correctly classified, resulting in high accuracy. This is a good indicator of the fact that the model could recognize satire even when composed of irony in most cases.

2. The misclassified samples were either specifically related to a political setting, or not explicitly satirical/ironic. It should be noted that the tweets in TWITTIRÒ were a few years older than the ones on which we trained the model, and thus the model likely had not seen references to certain individuals from the past.

3. Other examples were not recognized as satire because the humor was simply too subtle, or it required a deeper base of knowledge about specific aspects.

5.4.3.2 Missing Domain Knowledge

Consider this tweet which was misclassified by the model with high confidence (~87%) for the incorrect class:

(IT) *Schettino fa campagna elettorale per il Pd. Gli hanno dato i servizi sociali.*

(EN) *Schettino campaigns for PD. They gave him social services.*

In order to understand the tweet, one would need to know that Francesco Schettino was the captain of a capsized cruise ship *Costa Concordia*, who was sent to prison for manslaughter and abandoning the ship with passengers still on board. Thus, it is both impossible and absurd that he would be campaigning for a social services portfolio in the Italian parliament. Based on this example, one might conjecture that the model needs access to a knowledge base of some sort to classify texts that are tightly linked with the current political situation.

Thus, we investigated the performance on more recent tweets taken from the satirical feed, *twitter.com/Kotiomkin*. We used the model to predict a sample that refers to a quite recent event, in order to see if the knowledge base was better for recent tweets.

(IT) *Con Draghi la notte tra il 27 e il 28 potremo stare un'ora in meno in casa. Con Conte chissà cosa sarebbe successo #oralegale*

(EN) *With Draghi the night between 27 and 28 we will be able to stay at home an hour less. With Conte who knows what would have happened #legalhour*

This tweet refers to the fact that Draghi and Conte were recent prime ministers of Italy, and that Conte was known for having issued many presidential decrees restricting freedoms due to COVID-19 during his tenure. The tweet was correctly classified as satire, although the model had a very low confidence of 56%.

5.4.3.3 Use of Professional Language and the First Person
The samples that were written in a news-like style were not recognized as sarcasm, such as:

(IT) *Risultati provvisori elezioni tedesche: Spd 26,2% Cdu/Csu 24,7% Verdi 14,1% Fdp 11,5% Afd 10,6% Die Linke 5% Germania Viva 2%.*

(EN) *Provisional results of German elections: Spd 26.2% Cdu/Csu 24.7% Green 14.1% Fdp 11.5% Afd 10.6% Die Linke 5% Germany Viva 2%.*

The element of humor in this tweet comes from the insertion of the non-existent party "Germania Viva" with poor standing in the results of the German election. The aim here is to ridicule the real Italian party "Italia Viva." This tweet was misclassified by the model, with the most likely reason being that in our training set there were tweets from ANSA[4] that are written in a dry and precise style, like this one, and that make the satire more subtle and difficult to recognize.

(IT) *Bella e commovente telefonata di Draghi a tutti gli atleti medagliati: "Avete già pensato a come investire i soldi del premio?"*

(EN) *Beautiful and moving phone call from Draghi to all medalized athletes: "Have you already thought about how to invest the prize money?"*

The pun in this sentence is that Prime Minister Draghi was formerly the head of the European Central Bank. This previous tweet was also misclassified by the model. We chose to display it because it mimicked the style of another source in our training dataset, *Il Corriere della Sera*, and so the model interpreted it differently.

The model also confused tweets that are written in the first person. This could be explained by the fact that:

1. In our dataset, satirical sources are mostly written in the first person.

2. On the other hand, these tweets have no content that is usually found in that collection.

Therefore, the model struggles to balance these two reasons and predicts wrongly as a consequence.

After evaluating all the previous considerations, we can summarize the common characteristics of the tweets that the model mostly classifies as *Real*, when instead they are *Satirical*:

- They are short, concise, and dry.

- The knowledge base of the model should help in recognizing the theme of the tweet, so the tweet shouldn't be too new or too old.

- Sarcasm should not be too subtle, but quite explicit; otherwise, samples could be misclassified, or correctly classified with a very low confidence.

Moving on, we repeated this investigation with a source of real tweets, *la Repubblica*, and obtained similar observations:

- Tweets written in a style similar to tweets from satirical sources were mostly misclassified.

- Tweets outside the knowledge base of the model can be misclassified, with a very low confidence.

A portion of these results were similar to the ones we found in the satirical experiments.

To check if the conclusions we had drawn were correct, we took the worst predictions of the classifier and observed whether they could be put in one of these case studies; we didn't care about the language this time. This was also done to see if, in our limited knowledge about languages different from our mother tongue (Italian), the conclusion could be extended.

(EN) *Officially recognized as a national park in 1994, Joshua Tree obtained protected status after dedicated conservationists realized the area was in grave danger of hosting a future EDM festival.*

(EN) *Consumer Reports has rated the Onion Store the 1 place to buy Onion merch in America.*

The two English tweets above are dry and concise, and to know that they are satirical our model should be aware of what an EDM festival is (a dance festival), for the former, and what The Onion is (a satirical website), for the latter. Hence, this is exactly what we had supposed before.

5.4.3.4 Monolingual (English Only) Short-Text Classifier

We tested our English social media classifier on some of the most shared fake news in 2019 that we found on the internet. The results are shown in Table 5.4, and we can draw some conclusions about them.

- This kind of behavior can have the following explanations:

- In our dataset, fake news was represented by the titles of the *Fake* long texts. So, they were on average shorter than the tweets. This led the model to recognize shorter texts as an example of fake news. In fact, if we look at the third example in 4, it was longer and more detailed, and it was recognized as *Real*.

- The satirical sources used in this task for the English dataset were both of the English sources that we used in training: NewsThump and The Onion. None of these tweets were similar to their style, so none of them were recognized as satire.

- It is important to underline that, since these are all examples contemporary to the fake examples of our dataset (or almost all), the knowledge base of the model is sufficient.

TABLE 5.4   English Social Media Classifier, with Examples Taken from the Test Set

Text	Label	Predicted
Joe Biden Calls Trump Supporters Dregs of Society	Fake	Fake
NYC coroner who declared the death of Jeffrey Epstein a suicide made half-a-million dollars a year working for the Clinton Foundation until 2015	Fake	Real
Tim Allen quote Trump's wall costs less than the Obamacare website	Fake	Fake
Democrats Vote To Enhance Med Care for Illegals Now, Vote Down Vets Waiting 10 Years for Same Service	Fake	Fake
BREAKING: Nancy Pelosi's Son Was Exec At Gas Company That Did Business In Ukraine	Fake	Satire
Ilhan Omar Holding Secret Fundraisers With Islamic Groups Tied to Terror	Fake	Fake
Trump Is Now Trying To Get Mike Pence Impeached	Fake	Fake
AOC proposed a motorcycle ban	Fake	Satirical
Nancy Pelosi diverting Social Security money for the impeachment inquiry	Fake	Fake
Trump's grandfather was a pimp and tax evader; his father a member of the KKK	Fake	Real

We show the text next to the real label and the predicted one.

In addition, we performed an error analysis for an English-only model on the English test set. We used tweets, from *twitter.com/thedailymash*, that we had not used in training. It is interesting to highlight that we performed an error analysis on these same English tweets on the multilingual classifier, and we will see how the errors change, and to confirm furthermore that the conclusions of previous sections were good also for this language. In this analysis, we observed a few patterns that are worth mentioning:

- The Daily Mash and The Onion, that were in our training dataset, have a quite similar style. So, the majority of tweets were correctly recognized.

- When satire is too subtle, or not related to a political event in the knowledge base of the model, it is not recognized. An example of this is:

    (EN) *Couples lower their standards enough to marry each other.*

This was classified as *Fake*, and with a low confidence: ~38%. This means that the style of this sample was between *Satirical* and *Fake*, and the model got confused. On the other hand, in the multilingual model, this was correctly classified but with a very low confidence.

Another example of this is:

(EN) *Every female organism on earth to get divorced after hearing Adele's new song.*

Where the style is similar to the one used by CNN, even the tweets that were clearly *Satirical* are labeled as *Real*. In the multilingual model instead, this was classified as satire.

(EN) *Man shifts from gentle liberal to angry selfish b**** within seconds of getting in the car.*

Lastly, this one above was misclassified by both models.

- The model struggles when dealing with tweets written in the first person.

(EN) *I'm Anna, I'm happy to live in Milan.*

Using the different classifiers, these tweets were every time classified as *Satirical*.

5.4.3.5 Overfitting on Vocabulary

Analyzing erroneous predictions, we noticed that tweets regarding some particular people, or subjects, were classified by the model as ironic in most cases. We can see a clear example of a non-ironic tweet discussing the actions of the regional president of Liguria Giovanni Toti:

Toti fa gli auguri agli uomini che indossano la divisa della Polizia di Stato

(EN)*Toti greets the men wearing the uniform of the State Police*

Here removing the term "Toti" from the tweet results in the model's prediction changing from *satirical* to *real*:

fa gli auguri agli uomini che indossano la divisa della Polizia di Stato

(EN)*He/She greets the men wearing the uniform of the State Police*

The first tweet is not ironic, though it is classified as so, while the second was correctly classified. The word "Toti," as shown by the saliency map, pushes the model to make the wrong prediction. Another example of this issue is illustrated below. The following tweet which was invented by the authors as a synthetic example of a non-sarcastic tweet.

From this tweet we tried to cut out the most important word *difficoltà* and substitute it with *problema*. As can be seen, the sentence is predicted to be satirical:

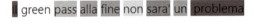

(EN) *The green pass will not be a problem in the end*

So, we made up another tweet:

(EN) *The mice in Rome: it will be a problem*

Where it can be easily seen that the word "problem" is probably causing the issue. So, it seems that the model recognizes irony based on the keywords alone.

For the long-text classifier, a suspicion came to mind, and we took one of the latest articles from The Onion, that was not in the training dataset:

> *LITTLE ROCK, AR—Noting the experienced hand with which she was able to put together a touching remembrance, family members confirmed Friday that area woman Dianne Melfi was getting pretty good at planning funerals. "At this point Mom is really in a groove when it comes to end of life arrangements—she's already done price comparisons of nearby funeral homes to get the best rate and she's memorized a half-dozen solid casserole recipes to feed the mourners with," said son Steve Melfi, 35, telling reporters that his mom was able to throw together a poignant slideshow of photos of the deceased with their family as though it was second nature.*

This was correctly classified, also with a very high confidence, almost 88%. How could this be? The explanation is simple. The Onion was the source most present in our dataset, so the model had learned well the style of this source.

Looking at the confidence matrix, we wanted to see if real news about a difficult topic (one about which satirical articles are often written) could be confused by the model. We tried this extract from CNN politics:

> *In Tuesday's elections, Republican candidates surged in blue states, cities rejected major police reform and suburban voters showed their independence. The major takeaways? This is a more moderate and centrist country than activists on either the right or left let on, and Donald Trump fever may be breaking. The system is working. Here's one thing everybody can be happy about: The election results, for the most part, are not being questioned. That may have a lot to do with Republicans doing well. But the results should prove to them that Trump's voter fraud myth is in fact a myth.*

The style of this text is quite similar to some sarcastic sources that we have in our training dataset, like The Onion, and moreover, it talks about Trump. So, the model classifies it as Fake, and it shouldn't.

5.4.4 Analyzing Saliency-Map–Based Explanations

We now discuss visual explanations produced using saliency maps, which associate to each input token an importance value in the range between 0 and 1. We show this importance by highlighting words in the sentence, where a darker color indicates higher importance. We make use of three colors to indicate the predicted class: green for real, red for fake, and blue for satirical (see Figure 5.3).

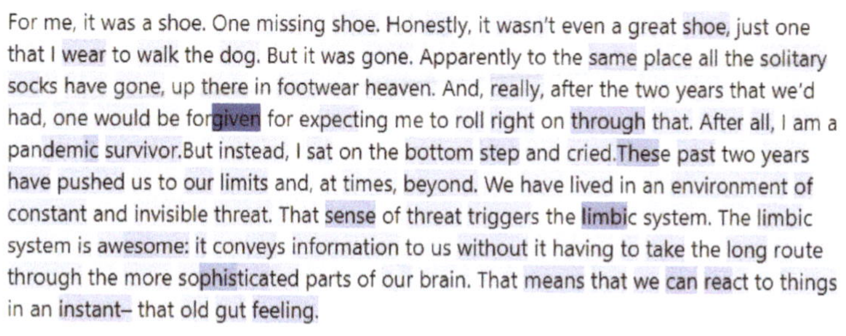

For me, it was a shoe. One missing shoe. Honestly, it wasn't even a great shoe, just one that I wear to walk the dog. But it was gone. Apparently to the same place all the solitary socks have gone, up there in footwear heaven. And, really, after the two years that we'd had, one would be forgiven for expecting me to roll right on through that. After all, I am a pandemic survivor.But instead, I sat on the bottom step and cried.These past two years have pushed us to our limits and, at times, beyond. We have lived in an environment of constant and invisible threat. That sense of threat triggers the limbic system. The limbic system is awesome: it conveys information to us without it having to take the long route through the more sophisticated parts of our brain. That means that we can react to things in an instant– that old gut feeling.

FIGURE 5.3   Saliency map for the long text.

For the latter question, we took from Snopes.com the link of this article from The Daily Exposé that was clearly fake:

> *Australians are currently living under one of the strictest dictatorial regimes in the world and now they are coming for the children. The Premier of New South Wales has acknowledged that without dramatically lower case numbers, even opening up at 80% vaccination rates will be difficult, as hopes of a lighter lockdown beyond August fade, and Brad Hazzard the Minister for Health and Medical Research has now told parents in a press conference that 24,000 children will be herded like cattle into a stadium to get the experimental Covid-19 vaccine, and parents will not allowed to be present.*

This was predicted as real. It is easy to spot why. The style is clear and concise, and the model is still out of its knowledge base. So, the model recognizes it as an example of its "real" set of samples and classifies it consequently.

5.4.4.1 Humor through Hyperbole, Antithesis, and Alliteration

In the following, we will look at some of the saliency maps we obtained using a gradient explanation method for the Italian tweets, hoping to find some patterns in words and some rhetorical figures to be recognized. They can be read as follows: the darker words are the most important for the prediction; as the color fades, the importance of the word diminishes. Looking at these maps, we found that some rhetorical figures are recognized more than others. As an example, in the first tweet we have a hyperbole:

> Aggiungo: multe a tutto spiano! Andremo a pile, sempre che# Super# Mario Monti non decida di tassare anche quelle!

(EN) *I add: full blown fines! We will go on batteries, unless #Super #Mario #Monti decides to tax those too!*

While in the second tweet, two words with a different sentiment together lead to a darker saliency map:

> se sento ancora la parola merito vomito# labuonascuola# chenonèquelladirenzi

(EN) *If I hear the word merit again I vomit #thegoodschool #thatitsnottheoneoffrenzi*

In the next tweet, the repetition of the letter 's' contributes to the prediction of the model. In particular, the first repetition of letter 's' are not so important, while the ones after the full stop '.' are:

Bambini scavano nella sabbia e trovano due chili di marijuana.
Ma solo se li addestri bene.

(EN) *Children dig in the sand and find two kilos of marijuana. But only if you train them well.*

In this other tweet instead, there is a particular kind of rhetorical figure that is called *antithesis*. In fact, the figure of Mario Monti, Italian economist and politician, is opposed to that of Chuck Norris, an actor. The interesting thing to see is that the word Mario Monti seems to be important for the prediction, and Chuck Norris, an actor who is ridiculed for being invincible in battle, not in the knowledge base (or not in the same measure as the former) is less important, but not completely insignificant.

Ma perche' scegliere Mario Monti quando ci sarebbe Chuck Norris?

(EN) *Why should we choose Mario Monti when we could have Chuck Norris?*

Another experiment we wanted to do was with a list of examples of rhetorical figures, in order to see which ones were recognized by the model and which were not. The first one here is the alliteration, repetition of the same sounds at the beginning or within words. An example was:

Far fuoco e fiamme

(EN) *Make fire and flames*

The repetition that can be seen here is of the letter "f"" As you can see, the more the model sees it, the more it becomes important. A figure that is very often used in satire is the anaphora, repeating the same word at the beginning of different parts of the sentence. We took a famous example, from *I Promessi Sposi*, a novel by Alessandro Manzoni:

Don Abbondio stava su una vecchia seggiola, ravvolto in una vecchia zimarra con in capo
una vecchia papalina.

(EN)*Don Abbondio was in an old chair, wrapped in an old cloak, with an old skullcap on his head.*

The first two "vecchia" words are less important than the last one, so this means that the repetitions are seen, but not right away, at least after the second word. Then, we analyze the "sister" of this rhetorical figure, the epiphora, in which the repetition was put at the end of different parts of the sentence.

Due persone che sentono di essere la stessa persona si amano, due persone che progettano lo stesso futuro si amano

(EN): *Two people who feel they are the same person love each other, two people who plan the same future love each other*

This also was not recognized as satire by the model. We consider it as a first hint of the conclusion of this section: if the style of the tweet is like the one we observed in the Error Analysis section, or the content is its knowledge base, then the rhetorical figure helps the prediction. Otherwise, that alone is not sufficient to make the model categorize the tweet as satire. This is an expected behavior.

Regarding the tweets we analyzed, during the exploration of data we saw that a lot of them have anaphora, so the one with Don Abbondio is explained. On the other hand, the word *"amano"* was evidently out of the knowledge of satire of the model, as the style of that text, and so it was classified as not satire. The alliteration is used in a lot of satirical tweets, and so the model classified them as satire.

5.4.4.2 Analyzing Part-of-Speech Tags

Another experiment that we performed is to look at the average explainability retained by each Part-of-Speech[5] (POS) in a sentence. To do so, we took our multilingual classifier, and we used it to predict a number of tweets from the Kotiomkin dataset and from TWITTIRÒ. Together with predicting and computing the explainability for each word, we used the POS tagging method available in Spacy[6] to get the POS tag of each word. We collected the results, computed the average importance of each tag, divided by correct and wrong classification, and the results are shown in Figures 5.4–5.6.

By looking at these figures, we first observe a difference in most important POS tags across the two sources, as ulterior evidence that the writing style matters in the detection and classification of satire. Then, another feature that appears from the figures is that while in the TWITTIRÒ dataset

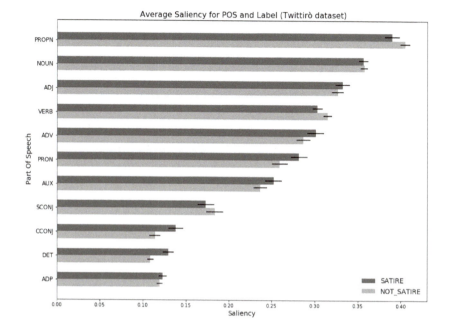

FIGURE 5.4   Average saliency values for each POS in TWITTIRÒ. The black bars indicate standard errors.

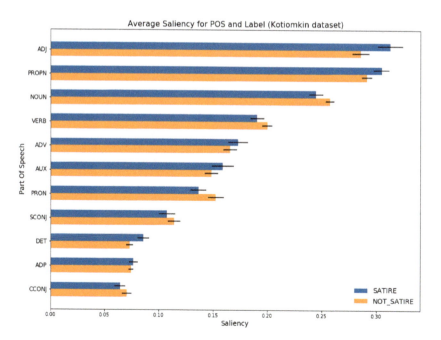

FIGURE 5.5   Average saliency values for each POS Kotiomkin datasets. The black bars indicate standard errors.

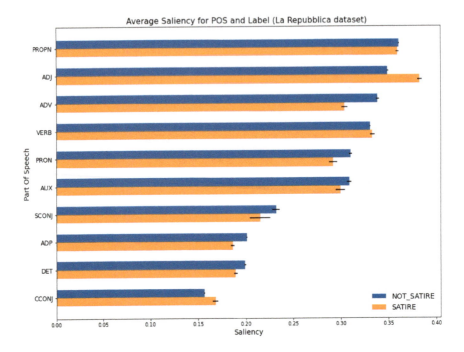

FIGURE 5.6 Average salience values for each POS in la *Repubblica* dataset. The black bars indicate standard errors.

the ranking of POS tags is similar both for correct and incorrect labels; in the Kotiomkin dataset, we observe a higher difference (highlighted in blue and orange).

We did the same analysis for some tweets in a real dataset, from Agenzia ANSA, and we observed the result in Figures 5.4–5.6. Some conclusions can be drawn after looking at all three images. As you can see these Figures, PROPN (proper name) and ADJ (adjective) seem to be quite important in predicting the Kotiomkin and TWITTIRÒ samples. On the other hand, in ANSA samples, it seems that NOUN and VERB play a more important role than PROPN: in fact, it seems that in this context PROPN gets the model confused. ADJ seems to be still important. So, in conclusion, the model seems to be learning that the structure: PRON-VERB-ADV is characteristic of REAL texts, while the other structure: ADJ-PROPN-NOUN is characteristic of SATIRICAL texts.

So, probably in our dataset the ironical tweets contained more proper nouns than the real ones, as it is easy to imagine. Also, it seems reasonable to think satirical tweets are written in a more personal way. Therefore, they are more foul-mouthed, ignoring the traditional structures of the sentence which, instead, would be more correct from a syntactic point of view, and that in fact is followed more by the real tweets, since we use official sources such as newspapers or news agencies.

5.5 CONCLUSIONS

We return to discuss the main research questions guiding our work. The first questions that we started from were:

1. Is it possible, using modern tools, to correctly classify short sentences as satirical?

2. Is it possible to build a classifier that is generalized enough so as to minimize errors in prediction and overcome the differences across languages and cultures?

3. Is the performance of a monolingual classifier better with respect to the performance of a multilingual one?

   These three questions can be answered together. In fact, with all of our experiments and study, we can definitely answer that as: yes, it is possible to correctly classify a short sentence as satirical, and this in a way answers the second question too. The problem is that overcoming cultural differences seems to be, after our work, slightly more difficult. One could also say that at least some knowledge about the language, or similar languages, that we want to classify must be included. Other samples were not recognized correctly, even if the language had been included in the training set. We think it might be because of their content, which wasn't contemporary (perhaps too old or too new) with respect to the content that the model had learned on. Therefore, as future directions, we suggest the following:

• Include the language that you want to perform the classification on at least in the training set, taking care that the samples are meaningful and significant.

• If this is not possible, or perhaps not wanted, choose some other language that is similar in syntax or in culture.

• Include samples that are old and new, in order to span the highest possible number of subjects and let the model learn well to generalize.

The next question to be analyzed is:

4. Making use of gradient-based methods, is it possible to produce an explanation that is right for the context and therefore build a tool that is human-interpretable and useful?

We adapted an existing method of explanation to better understand how our model worked, and the results were quite satisfying. Most of the previous works in classification similar to ours did not use the gradient method, but at the end we were satisfied with the results. Hence, we were able to answer positively to the question, that is, such methods can be used to better understand how the model works. As a future direction, one could include:

- Other methods for the explanation, still based on gradients, like the Integrated-Gradients method.

- Completely different methods, based on attention or neurons.

Once one does that, the different methods can be compared to see which one better captures the model learning process.

5. Is it possible, using modern tools, to correctly classify a text as fake, real, or satirical news?

6. Is it possible to display clearly how the prediction was made by the model and which parts of the sentences were responsible for it, in order to identify and see patterns in words and phrases and to analyze what linguistic features let the model make a mistake?

The fifth question moves us onto the second big direction of this work. The answer, as one could have understood at this point, is positive, but with some limitations. In fact, one big issue is related to the nature of fake news. Collecting fake news has been the real issue of this task. Thus, to make a study like this more significant and complete, one should rely on fact-checking researchers, or include in the dataset just news that have been classified as *Fake* not only by a single expert, or source, but by multiple ones, in order to be completely sure about the fact that this particular piece of news is unreliable, and so adopt more of a knowledge-based approach. Moreover, one could argue that someone, for political reasons, can label some texts as *Fake* when they actually aren't. This is outside the scope of this work, but our personal take is that it is high time that an external regulating organ (like the UN) rate the worthiness of news articles, due to their dangerous influence over populations and their extreme ease of dissemination.

Another issue is the fact that, very often, the styles of satirical and fake news overlap. As a future direction, this can be overcome easily by trying to select different kinds of satirical and fake articles, in order to include as many styles as possible in the training dataset and let the model learn better. Concerning the sixth and last question, in previous chapters we have shown that the saliency maps and the explainability metrics helped us in observing patterns in words and phrases. Style was shown to be important in this kind of analysis, and also some patterns, repetitions of words and rhetorical figures were correctly recognized by the model. However, one could implement an automatic tool that analyzes a high number of sources and selects the most confident prediction of the model, showing the saliency maps of this. Also, in this work we focused primarily on Italian tweets for searching word patterns: in order to have a more comprehensive analysis, one could repeat the same study procedure in the other languages, like English, Spanish, French, and German, to detect which patterns of which language are captured by the model, and then do the explainability analysis for each of these languages.

NOTES

1. https://www.kaggle.com/ekatra/fake-or-real-news/version/1
2. https://www.kdnuggets.com/2017/04/machine-learning-fake-news-accuracy.html
3. https://www.kaggle.com/clmentbisaillon/fake-and-real-news-dataset
4. http://www.ansa.it
5. Part-of-Speech (POS) tagging is the process of marking up a word in a text (corpus) as corresponding to a particular part of speech, based on both its definition and its context.
6. http://spacy.io

REFERENCES

1. Wikipedia contributors. Satire – wikipedia, the free encyclopedia. Online; accessed 23-10-2021.
2. Autorità per le Garanzie nelle Comunicazioni. Indagine conoscitiva su piattaforme digitali e sistema dell'informazione, 2021.
3. W. W. Chapman, and P. M. Nadkarni, Ohno-Machado L. Natural language processing: an introduction, Journal of the American Medical Informatics Association, 18(5):544–551, 2011.
4. Jacob Devlin, Ming-Wei Chang, Kenton Lee, and Kristina Toutanova. BERT: pre-training of deep bidirectional transformers for language understanding. *CoRR*, abs/1810.04805, 2018.
5. Ana-Cristina Rogoz, Mihaela Gaman, and Radu Tudor Ionescu. SaRoCo: Detecting satire in a novel Romanian corpus of news articles. abs/2105.06456:1074–1075, 2021.

6. Robert McHardy, Heike Adel, and Roman Klinger. Adversarial training for satire detection: Controlling for confounding variables. *Computing Research Repository, Computing Research Repository*, abs/1902.11145: 660–665, 2019.

7. Xinyi Zhou, and Reza Zafarani. Fake news: A survey of research, detection methods, and opportunities. *Computing Research Repository*, abs/1812.00315:137, 2018.

8. Bolei Zhou, Aditya Khosla, Àgata Lapedriza, Aude Oliva, and Antonio Torralba. Learning deep features for discriminative localization. *Computing Research Repository*, abs/1512.04150, 2015.

9. Marco Túlio Ribeiro, Sameer Singh, and Carlos Guestrin. Why should I trust you?: Ex-plaining the predictions of any classifier. *Computing Research Repository*, abs/1602.04938:1–10, 2016.

10. Ian E. Nielsen, Dimah Dera, Ghulam Rasool, Bouaynaya Nidhal, Prakash Ravi, and -machandran Ra. Robust explainability: A tutorial on gradient-based attribution methods for deep neural networks. *Computing Research Repository*, abs/2107.11400, 2021.

11. Wojciech Samek, Grégoire Montavon, Andrea Vedaldi, Lars Kai Hansen, and Klaus-Robert Müller. *Explainable AI: Interpreting, Explaining and Visualizing Deep Learning*. Springer, 2019.

12. Karen Simonyan, Andrea Vedaldi, and Andrew Zisserman. Deep inside convolutional networks: Visualising image classification models and saliency maps. *Computing Research Repository*, abs/1312.6034, 2014.

13. Jasmijn Bastings, and Katja Filippova. The elephant in the interpretability room: Why use attention as explanation when we have saliency methods? *Computing Research Repository*, abs/2010.05607:149–155, 2020.

14. Alec Radford, Jeff Wu, Rewon Child, David Luan, Dario Amodei, and Ilya Sutskever. Language models are unsupervised multitask learners. 2019.

15. Martín Abadi, Ashish Agarwal, Paul Barham, Eugene Brevdo, Zhifeng Chen, Craig Citro, Greg S. Corrado, Andy Davis, Jeffrey Dean, Matthieu Devin, Sanjay Ghemawat, Ian Goodfellow, Andrew Harp, Geoffrey Irving, Michael Isard, Yangqing Jia, Rafal Jozefow-icz, Lukasz Kaiser, Manjunath Kudlur, Josh Levenberg, Dandelion Mané, Rajat Monga, Sherry Moore, Derek Murray, Chris Olah, Mike Schuster, Jonathon Shlens, Benoit Steiner, Ilya Sutskever, Kunal Talwar, Paul Tucker, Vincent Vanhoucke, Vijay Vasudevan, Fer-nanda Viégas, Oriol Vinyals, Pete Warden, Martin Wattenberg, Martin Wicke, Yuan Yu, and Xiaoqiang Zheng. TensorFlow: Large-scale machine learning on heterogeneous systems, 2015. Software available from tensorflow.org.

16. J Alammar. Interfaces for explaining transformer language models, https://jalammar.github.io/explaining-transformers/, 2020.

17. B Gilbert. The 10 most-viewed fake-news stories on Facebook in 2019 were just revealed in a new report, https://www.businessinsider.com/most-viewed-fake-news-stories-shared-on-facebook-2019-2019-11?r=US&IR=T, 2021.

18. Alessandra Teresa Cignarella, Cristina Bosco, Viviana Patti, and Mirko Lai. Twittirò: An Italian twitter corpus with a multi-layered Annotation for irony. *IJCoL. Italian Journal of Computational Linguistics*, 4(4–2):25–43, 2018.

CHAPTER 6

Automated Techniques for Identifying Claims and Assisting Fact Checkers

Stefano Agresti and Mark J. Carman

Politecnico di Milano, Milan, Italy

6.1 INTRODUCTION

The phenomenon of *fake news*, involving the publishing of apparently real but actually false or misleading information, has been a plague on the Internet since the latter's creation. However, the worrying scale that the phenomenon has reached recently, combined with its mounting effects on the political discourse, is causing renewed concerns among the public[1]. Fighting this issue is, unfortunately, very difficult. With billions of posts and tweets shared online every day, fact checkers and journalists are at a disadvantage and it is becoming clear that, without the use of automated tools, it will be impossible to target online misinformation effectively. Fortunately, thanks to advances in the field of automated text classification, it is now conceivable to build systems capable of analyzing large quantities of text and automatically flagging those that may contain false or misleading information. In this chapter, we propose a new design and prototype for such a tool, providing empirical evidence supporting the approach, and presenting new strategies on how to employ Artificial Intelligence (AI) to detect fake news and assist fact checkers.

DOI: 10.1201/9781003296126-10

6.1.1 Beyond Fake vs. Real

Despite the exceptional achievements of the latest text classification technology and numerous studies produced on the subject of fake news detection (discussed below), we are still far from an effective and comprehensive solution for automatically detecting fake news. We argue that one of the main causes of this delay is the way researchers have been approaching the problem, focusing their attention on comparing "false news" against "real information." This strategy, common to data classification problems, lacks a holistic view of the phenomenon of fake news, since it disregards the subtle differences that exist between different types of fake news and the way they spread online. Thus, the first purpose of this chapter is to propose a new, more complex classification of online content, that goes beyond the binary distinction of "fake" vs. "real" news. We will then show that, using technology available today, it is possible to build an automatic classifier using such a taxonomy, presenting a prototype called *fastidiouscity*[2].

6.2 RELATED WORK

Before describing our contributions, we present an overview of the existing literature in the field of automated fake news detection.

6.2.1 Defining Fake News

When developing fake news detection systems, the first issue to address is to define an appropriate set of labels to classify the items. The concept of *fake news* is somewhat vague with experts and researchers yet to find a common agreement on its definition. Most papers have tried to deal with the problem by copying the approach of fact checkers, who define news based on how truthful their content is. This is usually implemented through the use of truth scales, or through a straightforward *true/false* approach. This is limiting, as it does not consider issues like misleading writing or biased reporting. Other papers, like Nakamura, Levy, and Wang (2019) and Molina et al. (2019), have tried classifying news based on why they are considered fake, for example using categories such as "misleading" or "imposter content." Another innovative approach, presented in Tandoc, Lim, and Ling (2017), used a two-dimensional classification, evaluating separately an article's factuality from its intention to deceive. Despite these efforts, none of the approaches has thus far achieved the goal of being both complex enough to capture most variations in fake news while at the same time simple enough to be employed in a classification context.

6.2.2 Fake News Detection

While there are many papers and researches focusing on how to fight fake news using AI, most of the proposed techniques can be divided into four categories:

- *Knowledge-based* techniques try to replicate the process of fact checking in an automated way. Their goal is to produce a system that can understand the content of a text, evaluating its truthfulness against a database of known information. This approach, while simple in theory, faces several practical limitations. Maintaining a constantly updated database of truthful information is extremely expensive, if feasible at all, and confronting it against thousands of texts automatically is complex even for modern technologies.

- *Style-based* techniques analyze how texts are written to determine whether they might be fake news. This can be coupled with the analysis of attached multimedia content in what is called *multimodal fake news detection*. This approach, to a certain extent, overcomes the issues of knowledge-based systems, and can be easily scaled, as it does not require expensive infrastructure to be maintained. However, its applications in real-world scenarios suffer from low accuracy and can be eluded by publishers who manipulate their writing style specifically to avoid detection.

- *Propagation-based* techniques assume, confirmed by various studies (e.g., Vosoughi, Roy, and Aral 2018), that fake and truthful news spread in fundamentally different ways. Many papers have proposed strategies to exploit this phenomenon, showing interesting results (e.g., Zhou and Zafarani 2019). The main drawback of this approach is that it is effective in recognizing fake news only after it has spread, giving it time to reach users and cause damage before it can be stopped.

- *Source-based* techniques focus on a news article's source, rather than its content, to determine its credibility. As shown in Horne, Norregaard, and Adali (2019), there is a clear distinction in how news publishers behave on social media, making it easy to detect suspicious sources. It is also effective to look at the users who spread a particular news, since fake news is often generated and spread by bots (Cai, Li, and Zeng 2017; Shao et al. 2018). Blocking malicious publishers and users could therefore be beneficial to the quality of online news content. However, marking news as *reliable/unreliable* depending solely on who created or shared the item could raise serious ethical concerns that should be evaluated before releasing a detection system.

For a more in-depth analysis of fake news detection techniques, we suggest Zhou and Zafarani (2018).

6.3 METHODS

We now describe the methods developed and used in our work, namely a new taxonomy for real and fake news and a multilayer classification model.

6.3.1 New Taxonomy

As explained in the previous sections, labelling online news is a complex task, balancing between the nuances of the field and the limits of technology. In this chapter, we propose a new taxonomy for online content that, by abandoning the one-dimensional approach in favour of a multi-layered one, splits the original problem of fake news detection into a series of smaller tasks, providing a classification that is both accurate and easy to implement.

The first layer of our taxonomy divides content into one of four categories, filtering *newsworthy* material:

- *News*: content of public interest, described in an objective manner.

- *Opinions*: content of public interest, but with the main purpose of expressing the writer's point of view on a subject.

- *Personal posts*: content that deals with people's private lives and is mostly inconsequential to the reader's perspective on the world.

- *Memes*: a *meme* can be defined as a multimedia content that has been modified in an evident way before being shared again. A *meme* is distinguishable from any kind of news content, but, precisely because of that, they have been shown to be an effective propaganda tool (Nieubuurt 2021).

The first layer of classification (Figure 6.1) allows us to discard most social media content, filtering out all *personal content*. In addition, since we wanted to focus on text (rather than image/video) classification

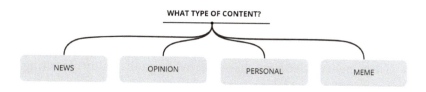

FIGURE 6.1   The first level of the classification.

technologies, we decided not to analyze further the *memes* category. Therefore, in the following steps we focus on the remaining categories: *news* and *opinions*, subdividing them as shown in Figures 6.2 and 6.3, respectively.

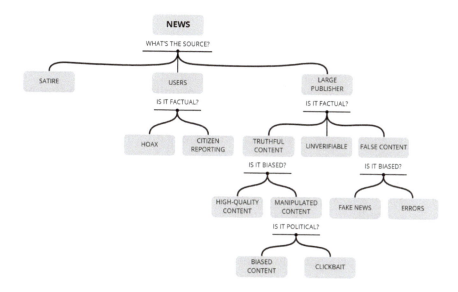

FIGURE 6.2   Classification of *news* in our taxonomy.

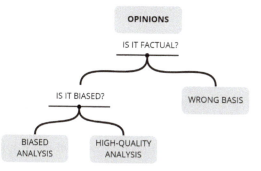

FIGURE 6.3   Classification of *opinions* in our taxonomy.

6.3.1.1 Subfield: News

We begin by dividing the news category based on the publishing source:

- *Large media publishers*: characterized by good quality of writing and large audiences. These sources are usually reliable.
- *Common users*: they have low following and relatively poor writing skills. They can report truthful news (*citizen reporting*), but they are mostly untrustworthy.
- *Satirical publishers*: their purpose is to mock the political establishment and are easy to recognize by the average reader.

The next division is based on factuality (satirical content is not included, since it is always non-factual):

- *Large media*:
  - *Truthful content*: news articles containing only verified (but not necessarily unbiased) information.
  - *False content*: news articles containing disproven information.
- *Common users*:
  - *Citizen reporting*: citizens reporting witnessed events on social media.
  - *Hoax*: false news in the form of fake citizen reporting, with somebody pretending to witness something that is untrue, or a conspiracy theory.

We then split *truthful* and *false content* based on whether the content is biased:

- *Truthful content*:
  - *Good quality content*: news reporting of only verified information, expressed in a generally objective manner.
  - *Manipulated content*: true stories portrayed in a way to favor a particular actor or to make the content more appealing to a particular group of readers.

- *False content:*

  - *Errors:* wrong information produced without any ill intention behind it.

  - *Fake content:* false content produced with the intention of misleading the reader.

Finally, we differentiate between the different ways a story can be manipulated, focusing on the writer's motivation:

- *Biased content:* stories that have been manipulated because of political motivations, usually through omissions, use of emotional language, or through an excessive emphasis on certain details.

- *Clickbait:* articles with the goal of drawing views to a website. Although less dangerous than biased content, clickbait is unethical and increases general distrust in news media.

6.3.1.2 Subfield: Opinion

For opinionated content, we adopted a simpler classification, focusing only on factuality and objectivity:

- *Opinions based on wrong information:* theses built on exaggerated or baseless claims. Readers should approach them with skepticism or discard them entirely.

- *Biased analysis:* articles where authors selected, analyzed, and presented the information available in a biased way, lowering the overall quality of their analysis.

- *Good quality analysis:* opinion pieces where authors, using correct information as a basis, provided a complete and objective analysis, only minimally influenced by their political stances.

It is worth noting that an opinion piece can be both biased and contain wrong information, although the latter is generally a more serious accusation.

6.3.2 Multilayer Content Classification

We show here how the categorization presented above can be implemented in an automated, multilayered classification system. The steps the system must perform are:

1. Determine content newsworthiness, filtering out personal posts.

2. Analyze the content source (only for news content).

3. Analyze the content factuality.

4. Analyze if the content is biased.

5. Analyze the author's intent (mainly for news content).

Based on the results of each of these analyses, the system will be able to assign any online content into one of the categories from the taxonomy discussed above.

6.3.2.1 Assisted Fact Checking

By utilizing existing technology, we were able to realize a simplified proto-type of the system just outlined, containing five layers: (1) a newsworthi-ness classifier, (2) a professionality classifier, (3) an automated fact-checking system, (4) a bias detector, and (5) a detector to evaluate the political ideol-ogy behind a text. For the third layer, arguably one of the most important in the system, we designed an *automated fact-checking system* that works as follows:

1. Given a text, it detects every *check-worthy* sentence contained in it.

2. For each of the claims, an online search is performed to find *related evidence*.

3. An *agreement detector* analyzes whether the retrieved evidence confirms or refutes the information contained in the original sentence.

With respect to the original structure, we simplified the last layer to fit the data at our disposal, moving from a more generic analysis of a writer's intention to a more specific predictor of what their political ideology could be.

6.4 DATASETS

This section provides a description of the datasets used in our experiments, divided according to the associated layer in the overall system.

6.4.1 Newsworthiness

For this layer we perform a three-way classification between news, opinions, and un- interesting content. To build a dataset for such classification, we exploited Reddit and its characteristic of dividing content according to monothematic communities, a strategy that we use multiple times throughout the chapter. To gather news articles, we collected 17,782 submissions from subreddit r/news[3], while for opinion pieces we scraped 15,816 articles from r/InTheNews[4]. As a collection of uninteresting content, we used an existing dataset from Kaggle[5], composed of blog posts divided by category. From this corpus, we removed posts labelled as politics or society and sampled 20,000 of the remaining entries.

6.4.2 Professionality

The main purpose of this layer is to discriminate between well-written articles and poorly written ones. As in the previous section, we exploited Reddit to build our dataset. To create a collection of low-quality articles, we scraped 11,688 articles described as clickbait or low-effort news from r/savedyouaclick[6]. Interestingly, among the top 5 publishers we found prominent news outlets, such as *Business Insider* and *CNN*, indicating that article quality may vary and/or sensational headlines may be present across all news agencies, not exclusively lesser-known ones.

To build the opposite dataset, given the multiple sources available, we decided to test three different strategies. First, we scraped articles from r/qualitynews[7], which provided 11,695 articles. Given the low number of subscribers of r/qualitynews, we also decided to test the dataset built from r/news presented in the previous section. This dataset is not only larger, but it is also more international than the one from r/qualitynews, although, being subject to less strict requirements, it might contain lower quality entries.

Finally, we created a third dataset by collecting news articles from seven specific newspapers renowned for the quality of their articles and in-depth analysis: *The Atlantic, Foreign Affairs, Politico, The New Yorker, The Economist, The Wall Street Journal*, and BBC. Across all of them, we obtained 15,437 unique samples.

6.4.3 Claim Detection

In order to develop the third layer of our system, the first step was to build a claim detection system. Different from the previous sections, we were able to find multiple papers on the subject, as well as two datasets.

The first dataset, provided by Hassan, Li, and Tremayne (2015), is composed by 20,000 sentences from debates, labelled as *check-worthy* or not. Since the labelling was performed manually, we argue that the dataset suffers from scalability issues.

The second dataset, presented in Atanasova et al. 2019, also consisted of sentences coming from debates, although in this case their label was based on whether they had been fact checked by *factcheck.org* or not. In our opinion, this approach presented numerous biases. Fact-checking organizations are more likely to fact check claims that are dubious (we point to Figure 6.4 for evidence), while they are less likely to focus on generic claims (examples in Table 6.1). In addition, fact checkers fact check a particular claim only once, even if it is repeated multiple times during a debate.

Thus we decided to introduce a new dataset using the following process:

- *Check-worthy* sentences were scraped from PolitiFact, collecting all the claims that have been fact checked on the website in the past ten years or so.

- As for the negative examples, we tried to imitate normal conversations by using the *Cornell Movie Dialogs Corpus*[8], a dataset of movie

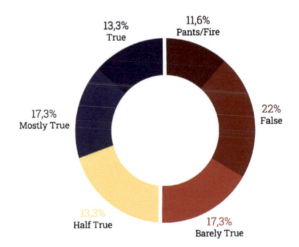

FIGURE 6.4 Distribution of scores across PolitiFact claims.

TABLE 6.1   An Extract from Atanasova et al. 2019

Sentence	Claim Label
So we're losing our good jobs, so many of them	0
When you look at what's happening in Mexico, a friend of mine who builds plants said it's the eighth wonder of the world	0
They're building some of the biggest plants anywhere in the world, some of the most sophisticated, some of the best plants	0
With the United States, as he said, not so much	0
So Ford is leaving	1
You see that, their small car division leaving	1
Thousands of jobs leaving Michigan, leaving Ohio	1

Reading the sentences, it is debatable whether some of them should be classified as claims or not.

scripts lines. To avoid introducing biases, we removed lines that were excessively long (more than 500 characters) or that were taken from sci-fi, fantasy, or historic movies.

We thus obtained 17,580 claims and 26,710 non-claims. In Figure 6.4 we show the ratings' distribution across the PolitiFact dataset.

6.4.4 Agreement Detection

To build a dataset for this task, we decided to use a collection of fact-checking articles with their fact-checked claim (truthful claims should agree with their fact-checking articles, while false claims should not). These articles were taken from PolitiFact and other news organizations found through the Google FactCheck API. The final dataset comprised of 52,877 entries, divided into 23 languages and 21 publishers.

Interestingly, observing the distribution of the entries by year and language (shown in Figure 6.5), we notice that the number of articles is

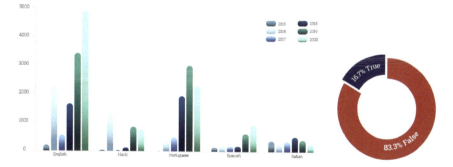

FIGURE 6.5   (left) The distribution of articles by year and language. (right) the percentage of supporting against refuting articles.

correlated to political events in different countries. Indeed, English articles have a spike in 2016, when US presidential elections were held, while Portuguese and Italian ones peaked respectively in 2019 and 2018, when Brazil's and Italy's general elections were held. Similar to the dataset from PolitiFact, the samples are characterized by a significant unbalance towards false claims, further reinforcing our speculation in Section 6.4.3 about fact-checking organization prioritizing suspicious claims over truthful ones.

6.4.5 Political Bias and Ideology

The final tasks, political bias and ideology detection, are strongly related, which is why we were able to use the same datasets in both cases. We will therefore present them here in one section.

The first dataset, specific to the bias detection task, was presented in Pryzant et al. (2020) and consists in 181,474 sentences taken from Wikipedia which didn't respect the website's *neutral point of view* policy. Each sentence comes with the edited, unbiased version.

For our purposes, we labelled the original sentences as *biased* and the edited ones as *unbiased* and dropped, for each entry, either the original or the edited sentence.

The second dataset is the *All the news* dataset on Kaggle[9], a collection of 2.7 million news articles which we rated based on their publishers' score on MediaBiasFactCheck. Since the resulting dataset was skewed towards left-leaning sources (with only one publisher being right-leaning), we decided to integrate it with 52,699 articles retrieved from subreddit *r/conservative*[10]. To balance it, we merged this data with an equal amount of left-leaning and unbiased entries from *All the news*. We used this dataset in both tasks, simply removing the unbiased articles for the ideology detection.

As mentioned before, we believe that rating an article based on its publisher is a questionable practice. Therefore, after building the previous dataset, we decided to expand the idea of using Reddit to retrieve left-leaning articles as well. By selecting left-leaning subreddits, we gathered 36,658 articles, which we then merged with the 52,699 coming from *r/conservative*. As unbiased articles, we selected 54,032 articles from *All the news* produced by *Reuters*, which we deemed authoritative enough for the purpose. As before, this dataset was used both for bias and ideology detection, simply discarding *unbiased* articles in the latter case.

Interestingly, if we look at the most used words in these two datasets, the word *Trump*, despite being the second most popular among *biased* articles, is not even in the top 30 when it comes to *unbiased* ones. This is probably a reflection of how news media used the figure of former US President Donald Trump to attract readers and viewers. Curiously, if we look at conservative and liberal articles, the former not only appear to be talking more about *Trump*, but they also talk more about *Biden* (respectively, first and tenth most used words).

6.5 EXPERIMENTS

In this section, we will discuss the experiments conducted in order to prove the validity of our system, as well as verify the quality of the datasets we built. We want to highlight that, being our main focus the creation of an environment to be used in future works, we decided not to test multiple models for our classifiers, but rather focused on experimenting with different strategies using BERT, which will be the base for all the models presented in this section.

6.5.1 Crowdsourced Evaluation of Dataset Quality

As we saw in the previous section, the most notable strategy we proposed for building new datasets is exploiting Reddit's peculiarity of creating monothematic communities (*subreddits*) to collect large amounts of posts or articles, labelled according to their original community. To test the reliability of this strategy, we set up a crowdsourcing experiment to evaluate whether human crowdworkers would agree with the labels we gave to the articles we collected. Showing them the submission from our ideology datasets, we asked them to decide if they were left- or right-leaning (each article was shown multiple times to lower the amount of noise). We received answers for 410 articles, of which 374 (91.2%) agreed with our labelling. Of the 36 erroneous answers, they were either *apolitical* articles (such as polls or similar) or they had been removed from the subreddits they came from. Only in a couple of cases the articles were self-criticism of one political side (such as an article denouncing a scandal surrounding Tulsi Gabbard on *r/democrats*)[11].

We believe that these results proved that Reddit can be a valuable source to build datasets of online content and that the quality of such datasets could be increased by discarding submissions with low or negative ratings.

6.5.2 Newsworthiness Classification

As explained in the previous sections, this classifier was designed to discriminate between *newsworthy* and uninteresting information, dividing texts into news, opinions, and personal posts.

TABLE 6.2   Classification Report for the First Newsworthiness
Classifier, Trained to Discriminate between All Three Categories at Once

	Precision	Recall	F_1-score
News	0.58	0.66	0.62
Opinion	0.57	0.46	0.51
Uninteresting	0.91	0.94	0.92

Its overall accuracy was 0.70.

TABLE 6.3   Classification Report for the Second Strategy

	Precision	Recall	F_1-score
Interesting	0.94	0.93	0.94
Uninteresting	0.96	0.97	0.96
News	0.60	0.61	0.60
Opinion	0.55	0.54	0.55

The overall accuracy of the first classifier was 0.95, while for the second
it was 0.58.

Using the datasets discussed in Section 6.4.1, we fine-tuned a BERT
model on them, obtaining the results shown in Table 6.2. Observing them,
we can infer that the model could easily distinguish between uninteresting
content from the rest, but it encountered more difficulties when deciding
between news and opinions.

Following the results from the previous experiment, we divided the task
into two subproblems, fine-tuning two different classifiers, one to detect
interesting content and one to classify it into news and opinions. Results
are reported in Table 6.3.

Unfortunately, this new strategy didn't improve but rather confirmed
the results of the first model. This showed that recognizing interesting
content is an easy task for BERT, while dividing it into news and opinions
is a harder problem to face. This is a satisfying result nonetheless, since
the most important task for this predictor was to "clean" the input of the
system, discarding irrelevant information.

6.5.3 Professionality Classification

The purpose of this classifier was to detect whenever an article suffered
from poor writing, an indicator that whoever wrote it is not a professional
journalist. To build this predictor, we compared BERT's performances on
three different combinations of the datasets presented in Section 6.4.2.
In all three cases, we used articles from *r/savedyouaclick* as low-quality
samples, while we varied high-quality ones. The first classifier used articles

TABLE 6.4   In Order: Classifier Trained on *r/qualitynews*, *r/news*, and Selected Publishers

	Precision	Recall	F_1-score
Low quality	0.91	0.88	0.89
High quality	0.88	0.91	0.90
Low quality	0.77	0.81	0.79
High quality	0.80	0.75	0.78
Low quality	0.83	0.81	0.82
High quality	0.86	0.88	0.87

TABLE 6.5   Comparison of the Three Different Classifiers

Datasets		Accuracy
Low quality	High quality	
r/savedyouaclick	*r/qualitynews*	0.89
r/savedyouaclick	*r/news*	0.78
r/savedyouaclick	selected publishers	0.85

collected from *r/qualitynews*, the second one used articles from *r/news*, and the last one used articles from selected publishers. Classification reports for all three classifiers are shown in Table 6.4.

In Table 6.5, we show the accuracy obtained by each classifier on its test set. Considering that we used the same model in all cases, it's safe to assume that their differences were closely related to how noisy each dataset was, which makes it unsurprising to discover that the worst-performing model was the one trained on *r/news*. It is more surprising to see that *r/qualitynews* proved to be a better source than the selected news publishers.

This might be an indication that, as we mentioned several times already, a news source is not always a reliable parameter to evaluate its quality. At the same time, this experiment shows again the validity of using Reddit to build accurate datasets.

6.5.4 Claim Detection

As part of our automated fact-checking system, we needed to develop a classifier to detect *check-worthy* claims inside a text. We described in Section 6.4.3 our approach to building a scalable dataset of *check-worthy* claims and normal sentences. The BERT model we trained on this data obtained an accuracy close to 1.00, showing that our sources produced clearly distinguishable categories. To test our classifier's performances on real-world data, we replicated the work of Hassan, Li, and Tremayne (2015), building a dataset of sentences labelled through crowdsourcing. To use a real-world

TABLE 6.6   Results Obtained by Our Claim Detector

	Precision	Recall	F_1-score
Not Claim	0.72	0.78	0.75
Claim	0.63	0.55	0.59
Not Claim	0.76	0.80	0.78
Claim	0.66	0.60	0.63

On the top, results on the entire test data (overall accuracy 0.69); on the bottom, results on test data limited to sentences reviewed at least twice (overall accuracy 0.72)

scenario, we used 4,018 sentences from the 2020 US presidential debates. On the 2,421 sentences labelled by crowdworkers, our model obtained the results shown in Table 6.6. The table also shows the results obtained on the subset of sentences labelled by at least two crowdworkers (little more than 950). In this case, performances were slightly better.

Final accuracy was 0.69, comparable to the best performance obtained by Hassan, Li, and Tremayne (2015) of 0.70. In that paper, however, the models were trained and tested on the same data and showed unbalanced results, with high precision and low recall. Our model shows a more balanced performance across the two categories, thus reducing the risk of overfitting in real-world scenarios. We believe that our experiment suggests that our approach for building datasets for this task is an effective one. Not only that, it has large room for improvements. Our collection of claims can be expanded with lower effort, even to new languages, while negative examples can be improved by using more variegated sources. Moreover, refining the testing data by continuing the crowdsourcing experiment might help in reducing noise and getting more accurate results.

6.5.5 Claim Reformulation

After detecting a claim, our system searches online for evidence that either supports or refutes it. While working on this step, we realized that often sentences were difficult to understand when extracted from their text (e.g., *He said that* is meaningless if not framed). Since this issue would affect the overall quality of the system, we decided to tackle it. Using an approach similar to Suresb (2020), we used *spaCy* [12] to detect all pronouns and entities in a given text. Each of the pronouns thus found was in turn substituted with the special BERT token *[MASK]*, before performing masked word prediction. The predicted word was chosen among the entities found in the text, and the prediction was only accepted if the model surpassed a given threshold of confidence.

We tested our approach on the GAP dataset from Webster et al. (2018), the benchmark for tasks of coreference resolution. We limited our test set to 286 sentences where the pronoun is either "He" or "She," discarding other pronouns for which our model had not been adapted. On these, our system reached an accuracy of 0.75, beating the baseline presented in Webster et al. (2018) of 0.66 and similar to the accuracy of 0.76 obtained in Suresb (2020).

Although the performances are likely to degrade in a real-world use case, it is noteworthy that the model we used wasn't even fine-tuned for the task (BERT can perform masked word prediction out of the box). It's therefore plausible that with an appropriate fine-tuning process these results might be improved, showing that this is a promising approach to the problem.

6.5.6 Agreement Detection

The last step of the fact-checking system is to analyze whether the evidence found online supports or refutes the initial claim.

For this task, we presented in Section 6.4.4 a dataset of 52,644 fact-checking articles from around the world, each accompanied by the related claim and truth rating. By training a BERT model on this data, we wanted to achieve a model that, given a sentence and an article connected to it, would be able to define whether the latter agreed with the former, or vice versa. Since the dataset was heavily skewed towards false claims, we trained the model in three different settings:

- Using the original dataset, without any changes.
- Using a balanced version of the dataset, sampling the "false" pairs.
- Using only the titles of the fact-checking articles from the balanced dataset.

The comparisons between the results from the various models can be seen in Table 6.7. Looking at performances, the last two models appeared to

TABLE 6.7 Comparison between Accuracy Values of Different Agreement Detectors

Model	Accuracy
Base dataset	0.82
Balanced dataset	0.68
Title only	0.68

The higher accuracy on the first dataset is misleading, as it was obtained by simply labelling the majority of samples as "false."

be almost equivalent, with a 0.68 overall accuracy in both cases, while the first ended up overfitting and labelling most of the test rows as false. After discarding this model, more experiments should be conducted to assess whether any statistical difference exists among the other two; although for our system we decided to use the one trained on the entire articles, as it would be easier to deploy.

6.5.7 Bias Detection

In our taxonomy, we established that a key role in determining the quality of a news article would be its level of objectivity. Using the datasets presented in Section 6.4.5, we trained three BERT models for this goal.

The first classifier was trained over the dataset of Wikipedia sentences from Pryzant et al. (2020), but its results turned out to be inconclusive, with the model labelling almost all samples in the test data as unbiased. The second classifier was trained on a dataset of our creation, built combining articles from the *All the news* dataset with articles from *r/conservative*. Although the final dataset was considerably skewed towards biased samples (with a ratio of 65/35), results on the test data were close to 100% accuracy, correctly identifying 33,409 articles out of 33,573. We then trained a third classifier on a dataset composed of unbiased articles from *Reuters* and biased articles from liberal and conservative subreddits. In this case too, the dataset was skewed towards biased entries, yet the model still got an accuracy close to 100%, even outperforming the previous one (28,601 correct predictions out of 28,676).

As with other classifiers we trained, we shouldn't be expecting these performances in real-world scenarios. The data still suffered from some limitations, especially with regards to unbiased articles. Moreover, since we're using test and training data coming from the same source, these results are inevitably more optimistic than they should be. Despite these issues, however, our experiments proved that BERT is capable of handling this task and supported our belief that Reddit can be used to create reliable news datasets.

6.5.8 Political Ideology Detection

For this last layer, we developed a classifier to determine an article's political stance. Given the close connection between this layer and the previous one, this classifier was built using the last two datasets from the previous layer, simply dropping unbiased articles and using as categories left-against right-leaning.

However, while in the previous case, the two datasets gave similar performances, for this layer the first dataset (which combined articles from *All the news* and *r/conservative*) produced inconclusive results, with the final model labelling most test data as right-leaning. The second classifier, trained on the Reddit dataset, performed instead much better, with an overall accuracy of 0.90.

These results showed that this is yet another task that BERT can handle effectively and were a further proof of the quality of the data extracted from Reddit, which is particularly fit for political-related research given its natural tendency to create closed and polarized communities.

6.6 ASSISTED FACT-CHECKING PROTOTYPE

While completing our experiments, we used the classifiers that we trained to build a prototype for a real-world use case of our research. This took the form of a web application called *fastidiouscity* (a screenshot is shown in Figure 6.6). The application, given a text, returns a series of analysis on the text bias, ideology and professionality, as well as analyzes each sentence of the text, evaluating whether it's a claim, searching for evidence online through the system we described in Section 6.5.5.

While still too unreliable to become a completely automated fact-checking system, we believe it could be extremely helpful as an *assisted* fact-checking system, helping fact checkers to speed up their work.

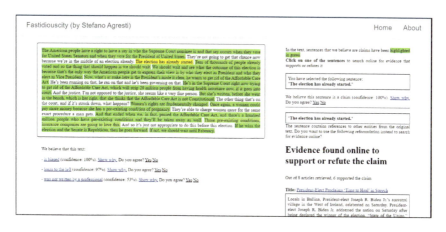

FIGURE 6.6　A screenshot taken from *fastidiouscity*, our working prototype. The text, extracted from the 2020 US presidential debate, was pronounced by then Democratic nominee Joe Biden.

Nevertheless, such a product would be a remarkable milestone, since the main drawback of traditional fact checking is its slowness compared to that of fake news.

6.7 CONCLUSIONS

In this chapter, we discussed the problem of online content classification, with the ultimate goal of building a tool capable of discriminating between reliable and unreliable information.

In Section 6.3, we introduced a new taxonomy that, by classifying online content through a multidimensional approach, managed to capture the different ways in which information can be manipulated, without requiring unrealistic performances from technology. Based on this classification, we built a working prototype, showing that it's feasible to build a tool to assist fact checkers using today's AI technology.

We tackled the lack of publicly available, high-quality datasets, showing a new strategy to build them exploiting Reddit. Such strategy was tested in Section 6.5.1 through a crowdsourcing experiment, where human crowdworkers confirmed the correctness of our labelling in over 90% of the cases.

Lastly, in Section 6.5.4 we introduced a new scalable dataset for the claim detection problem. We showed that the BERT model trained on test data labelled by human crowdworkers obtained results comparable to the baselines indicated in the literature.

6.7.1 Future Works

It's unlikely that *fake news* and misinformation will disappear in the near future. On the contrary, the problem of fake news detection will become only more and more relevant, with a particular emphasis on AI-powered solutions. We believe that each of the points we tackled in this chapter can be improved through deeper investigation and greater resources. In the following list, we outline some of the points we consider more important:

- Increase dataset size, by repeating our experiments on a larger scale.
- Extend to multimodal classification.
- Introduce a satire detector.
- Introduce a hoax detector.
- Conduct real-time analysis, by pairing our system with a voice-to-text system.
- Conduct qualitative analysis with journalists.

NOTES

1. https://time.com/5887437/conspiracy-theories-2020-election/
2. So named because the system would be considered fastidious by those spreading false claims and misinformation.
3. https://www.reddit.com/r/news/
4. https://www.reddit.com/r/inthenews/
5. https://www.kaggle.com/rtatman/blog-authorship-corpus
6. https://www.reddit.com/r/savedyouaclick
7. https://www.reddit.com/r/qualitynews/
8. https://www.cs.cornell.edu/~cristian/Cornell_Movie-Dialogs_Corpus.html
9. https://www.kaggle.com/snapcrack/all-the-news
10. https://www.reddit.com/r/conservative
11. https://www.reddit.com/r/democrats/comments/eetyvs/former_hawaii_governor_calls_on_tulsi_gabbard_to/
12. https://spacy.io/

REFERENCES

Atanasova, P. et al. (2019). "Overview of the CLEF-2019 CheckThat! Lab: Automatic Identification and Verification of Claims. Task 1: Check-Worthiness."

Cai, C., L. Li, and D. Zeng (2017). "Detecting Social Bots by Jointly Modeling Deep Behavior and Content Information." In: *Proceedings of the 2017 ACM on Conference on Information and Knowledge Management (CIKM '17)*. Association for Computing Machinery, New York, NY, USA, 1995–1998. https://doi.org/10.1145/3132847.3133050

Hassan, N., C. Li, and M. Tremayne (Oct. 2015). "Detecting Check-Worthy Factual Claims in Presidential Debates." In: *Proceedings of the 24th ACM International on Conference on Information and Knowledge Management (CIKM '15)*. Association for Computing Machinery, New York, NY, USA, 1835–1838. https://doi.org/10.1145/2806416.2806652

Horne, B. D., J. Norregaard, and S. Adali (2019). Different Spirals of Sameness: A Study of Content Sharing in Mainstream and Alternative Media. arXiv: 1904.01534 [cs.CY].

Molina, M. et al. (Oct. 2019). ""Fake News" Is Not Simply False Information: A Concept Explication and Taxonomy of Online Content." In: *American Behavioral Scientist*, doi: 10.1177/0002764219878224.

Nakamura, K., S. Levy, and W. Y. Wang (2019). "r/Fakeddit: A New Multimodal Benchmark Dataset for Fine-Grained Fake News Detection." In: *arXiv preprint arXiv:1911.03854*.

Nieubuurt, J. T. (2021). In: *Frontiers in Communication*. https://www.frontiersin.org/articles/10.3389/fcomm.2020.547065/full#h8.

Pryzant, R. et al. (Apr. 2020). "Automatically Neutralizing Subjective Bias in Text." In: *Proceedings of the AAAI Conference on Artificial Intelligence* 34.01, pp. 480–489 doi: 10.1609/aaai.v34i01.5385.

Shao, C. et al. (Nov. 2018). "The Spread of Low-Credibility Content by Social Bots." In: *Nature Communications* 9.1. doi: 10.1038/s41467-018-06930-7.

Suresb, A. (2020). "BERT for Coreference Resolution.".

Tandoc, E., Z. Lim, and R. Ling (Aug. 2017). "Defining "Fake News": A Typology of Scholarly Definitions." In: *Digital Journalism* 6, pp. 1–17. doi: 10.1080/21670811.2017.1360143.

Vosoughi, S., D. Roy, and S. Aral (2018). "The Spread of True and False News Online." In: *Science* 359.6380, pp. 1146–1151. doi: 10.1126/science.aap9559.

Webster, K. et al. (2018). "Mind the GAP: A Balanced Corpus of Gendered Ambiguous." In: *Transactions of the ACL*, to appear.

Zhou, X. and R. Zafarani (Nov. 2019). "Network-based Fake News Detection." In: *ACM SIGKDD Explorations Newsletter* 21.2, pp. 48–60. issn: 1931–0153. doi: 10.1145/3373464.3373473.

Zhou, X. and R. Zafarani (2020). "A Survey of Fake News: Fundamental Theories, Detection Methods, and Opportunities." In: ACM Computing Surveys 53.5, pp. 1–40, Article 109 (September 2021). https://doi.org/10.1145/3395046

V

Applications in Healthcare

Whisper Restoration Combining Real- and Source-Model Filtered Speech for Clinical and Forensic Applications

Francesco Roberto Dani, Sonia Cenceschi,
Alice Albanesi, Elisa Colletti,
and Alessandro Trivilini

Servizio di Informatica Forense – Scuola Universitaria Professionale della Svizzera Italiana, Lugano-Viganello, Switzerland

7.1 INTRODUCTION

This chapter focuses on the structure and development of the algorithmic components of VocalHUM, a smart system aiming to enhance the intelligibility of patients' whispered speech in real time, based on audio only.

VocalHUM aims to minimize the muscular and respiratory effort necessary to achieve adequate voice intelligibility and the physical movements required to speak at a normal intensity. It is primarily designed for patients in a temporary or prolonged state of physical and vocal frailty (e.g., respiratory infection, geriatric weakness, or partial/total paralysis) with the ultimate purpose to facilitate the patient–caregiver communication and improve the use of speech-to-text tool and voice commands. However, the

DOI: 10.1201/9781003296126-12

real-time enhancing algorithm, or HUM, focuses on speech characterized by correct articulation and proper linguistic production, in the absence of speech disorders such as apraxia or dysarthria. Indeed, we consider unvoiced speech restoration and enhancement as a first necessary step to move forward the state of the art for future solutions involving the linguistic components of the speech. The VocalHUM system consists of a small and smart object composed of a:

- Dynamic microphone positioned on the cheek near the lips.
- Single board computer: a tiny self-embedded headless computer.
- Wearable microphone stand.
- Built-in speaker used to reproduce the output.

HUM can be considered the "heart" of the VocalHUM technology and runs on the single board computer. The HUM structure and configuration have been fixed after a series of tests that consider different approaches. In particular, this chapter presents and discusses the choices that make us move from neural networks to a generative approach, addressing each issue for an application in a real context perspective.

To date, HUM is focused on the Italian language and combines real whispered speech, synthetized vowels (generated by means of additive synthesis) and consonant enhancement techniques. This approach provided the first promising results on the whisper of able-bodied speakers, and it is currently being tested on patients in real contexts. In summary, the presented work deepens the development and implementation path of HUM, with the following structure. Section 7.2 reports a summary of the state of the art regarding the characteristics of the whispered speech, its differences with respect to normal speech, and related existing speech assistive technologies in clinical contexts. Section 7.3 describes HUM's development history and the choices that lead us to abandon the machine learning techniques in favor of a generative approach. Section 7.4 describes HUM's architecture in its current version, while the following sections are dedicated to results, future works, and discussion.

7.2 THE STATE OF THE ART

Speech intelligibility and reduced mobility are common issues in hospital settings and represent a central interesting topic for many other applicative fields too. However, there is still a lack of commercial products and marketed solutions implementing the state of the art regarding real-time processing algorithms for speech enhancement and reconstruction. Due to this reason, the state of the art is structured in different sections.

The first is a fundamental overview to understand the involved acoustic features and behaviors regarding whispered speech and its differences from normally phonated speech. The second one is a summarized state-of-the-art regarding the communication issues of patients in real contexts, while the third focuses on Whisper-to-Speech (WTS) and enhancement prototypes, algorithms, and research. The last shortly introduces the added value of audio forensic techniques in facing speech reconstruction challenges.

These topics are fundamental in order to contextualize both the challenges we had to face, and in particular the real-time necessary processing, the space and power supply potential of the apparatus, and the suitable technology to allow the prototype to be worn and used by end users.

7.2.1 Whispered Speech

HUM applies a Whisper-to-Speech (WTS) transformation relying only on the audio streaming, without considering the linguistic component of the speech. This choice derives from the need to not burden the process not depending on phone models and annotated databases. A typical approach to speech enhancement exploits, for example, IPA annotations (Kallail & Emanuel, 1984; Robinson et al., 1994; Riekhakaynen, 2020; Strassel et al., 2003;), but these annotated corpora are extremely rare for whispered speech and, as far as we know, are not available for the Italian language. At the same time, it must be emphasized that whisper enhancement relying on the audio signal alone is a challenge not to be underestimated.

For example, increasing the intensity of the signal is not enough to enhance intelligibility especially in real contexts, and it is rather necessary to reconstruct the missing or attenuated language-related segmental and suprasegmental components of the speech (French & Steinberg, 1947; Amano-Kusumoto & Hosom, 2011). We refer here to a whispered speech belonging to the so-called soft whisper group (Weitzman et al., 1976). We do not describe here the physiology of this speech modality, which can be delved into in Morris & Clements (2002), Tsunoda et al. (1997), or Lim (2011).

Its main acoustics characteristics derive from the lack of vocal-fold vibrations which implies the absence of fundamental pitch, and the derived harmonic relationships among formants (Jovičić, 1998). Then, with respect to normal speech, the whispered speech presents several differences (Gao, 2003; Morris & Clements, 2002; Sharifzadeh et al., 2010) such as:

- Almost a total lack of F0 (and related features).
- Much flatter power-frequency distribution of formants between 500 and 2000 Hz.
- Formants' values tend to be higher in frequency.
- F1 usually shows the greater increase.
- F2 tends to be as powerful as F1.
- It has normally 20 dB lower power than its phonated version.
- Longer lasting vowels.

Another crucial point is the perceptual aspect: whispered speech is commonly understood by the human auditory system in optimal conditions. Perceptive issues start to present when the conditions are not exactly optimal, such as the presence of environmental noise. As underlined by Loizou and Kim (2010), despite progress in the development of speech enhancement algorithms focusing on speech quality, little progress has been made in improving speech intelligibility in critical conditions.

Moreover, it should be noted that phone calls reduce the useful speech bandwidth with repercussions on voice quality, while a traditional audio signal amplifier in many cases does not increase intelligibility, which depends on cross-segmental and suprasegmental factors of speech (Bradlow et al., 1996). Then, the identification and reconstruction of the acoustic features which vary between whisper and normal speech seem to be the first problems to be solved in order to improve the state of the art.

7.2.2 Speech Assistive Technologies in Clinical Contexts

The COVID-19 pandemic, which started in 2019, brought to light preexisting communication and isolation issues in clinical contexts (Sun et al., 2021). These contexts are equipped (in the best of cases) with internal phones or emergency bells, but nothing specific is adopted for speech intelligibility enhancement or restoration. A sectorial investigation highlighted the presence of various patented methodologies and tools for speech enhancement, not strictly designed for the healthcare field. However, we're speaking of algorithms or circuits that are still limited to noise reduction, intensity gain, equalization or echo cancellation[1], and therefore not strictly focused on speech (Berkovitz, 1982; Godsill et al., 2002). Some industries sell speech enhancement tools, but these are only focused on tasks like noise removal and acoustic echo cancellation finalized to improve speech recognition, which is beyond our scope. Main examples are Cerence[2], Alango[3], and Advanced Bionics[4].

More of the recent research attempts are focused on partially or totally laryngectomized patients, which are currently constrained to the use of invasive tools, laryngophones, or throat microphones (Cohen et al., 1984; Erzin, 2009; Liua et al., 2007; Sahidullah et al., 2016; Shahina & Yegnanarayana, 2007; Zheng et al., 2003). When they learn to speak autonomously, they use the so-called esophageal speech, which is much more difficult to understand than whispered speech due to the partial or total lack of vocal cords and the different phonation modality.

The available solutions for this target are yet not acceptable to reach a large-scale success among patients. Hardware devices are still limited to old-fashioned tools like an artificial larynx and a simple signal amplifier. Moreover, companies involved in this field do not seem interested in implementing the current speech enhancement techniques to their products; instead, they are applied to other domains. Some examples are Luminaud[5], Griffin Laboratories[6], and Atos Medical[7]. UltraVoice[8] is another available device in the market which is still very invasive. In summary, in a first step, HUM aims to reconstruct whispered speech of a healthy speech apparatus, but nothing prevents improving it in the future to also cover esophageal speech.

7.2.3 Speech Enhancement and Reconstruction Techniques

Speech intelligibility is a topic of transversal interest for human activities. It plays a substantial role in digital audio forensics in perception of linguistic contexts in environmental and telephone interception. It also applies to countless areas, such as home automation (e.g., voice commands in human–computer interaction), public data, and behavioral security (e.g., emergency voice commands on aircraft), and hardware and software applications for sung and spoken voice for artistic performances. Machine Learning (ML) and signal processing research has enormously improved in recent years. However, scientific research has rarely turned into usable products for the speech in clinical contexts. This is probably due to the vastness of the topic and its wide possible application joined with the relative novelty of many techniques.

Many signal processing techniques demonstrate the possibility of enhancing speech intelligibility, but they need to be refined and strengthened to be used in everyday life. Nevertheless, none of the newly developed algorithms appear to be yet implemented in products for real-time usage. Real-time processing is extremely fundamental for concrete exploitation, but it requires experimentation in real contexts, well-defined speech databases, an interdisciplinary approach, and most importantly constant interfacing with end users.

Recent advancements in Artificial Intelligence focused on speech enhancement techniques, successfully increased the intelligibility of Non-Audible Murmur (NAM) microphones by means of Generative Adversarial Networks (GANs) applied to NAM-to-Whisper (NTW) and Whisper-to-Speech (WTS) tasks (Patel et al., 2019; Pascual et al., 2019a; Shah & Patil, 2020; Zhou et al., 2012). Other techniques make use of Hidden Markov Models and Gaussian Mixture Models to perform speaker recognition (Patel et al., 2019) and speech enhancement (Doi et al., 2010). Nevertheless, none of these algorithms has been implemented in products for real-time usage.

The problem with GANs is that they are significantly difficult to train in terms of computational complexity (Patel et al., 2019). Even with these computational problems, these kinds of neural networks can produce significant results. In GAN architectures for WTS tasks, a GAN is trained with power spectrum and different spectral features to produce speech from a NAM signal, by finding the relation of the respective feature vectors (Toda et al., 2012). In the case of a Deep Neural Network, the procedure is almost the same but requires a higher number of hidden layers and, thus, allows to learn more complex relations between source and target spectral feature vectors, with a consequent higher computational cost and temporal delay.

Gaussian Mixture Models (GMMs) were successfully applied in tasks such as Voice Conversion (Stylianou, 1996). They lead to successful results also in WTS tasks if combined with Maximum Likelihood Estimation (MLE), as stated in Toda & Shikano (2005). In GMM architectures for WTS tasks, pairs of NAM and speech signals are fed to two GMMs for spectral estimation and pitch estimation. In this case, the features and the power spectrum of both signals are previously extracted (Toda & Shikano, 2005), and the model tries to find the mapping between the source and target feature vectors. In the approach exploiting Hidden Markov Model (HMM), the source and target parameters are modeled by a context-dependent phone-sized HMM. Then, an HMM recognition is applied on the input feature vectors, and, lastly, a third HMM is used for the synthesis of the speech (Tran et al., 2009).

Seeing the magnitude of the state of the art, at first we decided to approach the challenge based on ML techniques already mentioned in literature, but the result was unsatisfactory from several points of view (Section 7.3). We therefore opted for a generative approach (Section 7.4), which has been refined with mixed techniques crossing the original whispered speech with synthetic vowels in order to preserve the patient's timbre.

7.2.4 Added Value of Digital Audio Forensics

Digital audio forensic techniques play a pivotal role in the exploitation of forensic noisy and degraded speech recordings for solving a wide range of questions. Therefore, it can be extremely helpful for facing applied experimental research affected by a great amount of potential technical risks and subjective/objective variability, such as in clinical applications. These methodologies concern both the analysis of linguistic contents and the extraction of environmental and context noises, but most of the digital audio forensic applications are focused on human speech; for example, a speaker's timbre characterization, speaker recognition, speaker profiling (age/gender/geographical origin), or characterization of pronunciation defects in relation to a linguistic norm (see Broeders, 2001; Olsson, 2018 for an introduction).

Moreover, forensic speech analysis has a strong interdisciplinary nature, combining different linguistic competences such as signal processing, audio coding and compression, digital restoration, psychoacoustics, perception, and linguistics (especially phonetics, phonology, and sociophonetics). In forensics, both the segmental (vowels, consonants, silences, etc.) and the suprasegmental (the prosodic temporal behavior of local features such as intonation, amplitude envelope, etc.) levels and their mutual relationship are considered.

Signal processing and feature extraction techniques are constantly improved and, at the same time, the sociolinguistic and perceptive themes are considered, making the approach interesting for the clinical field, too. Finally, compared to traditional linguistic and phonetic sciences, digital audio forensic techniques offer a different multidisciplinary and concrete approach.

7.3 HUM's DEVELOPMENT HISTORY

HUM is designed for a speaker with healthy vocal chords emitting whispered or extremely faint speech due to different pathologies. This choice will allow to gradually test and refine HUM on real patient data and then solidly move towards new challenges such as, for example, esophageal speech. We started from the real requirements to find the most suitable algorithmic solution based on the following needs:

- Relying on the audio component only without exploiting the linguistic one.

- Making sure the overall WTS process on continuous audio streaming works with the lowest possible latency.

- Making sure the HUM algorithmic component can work in a small, wearable, and smart object.

The initial choice was to rely on previous promising research involving the use of ML techniques. In terms of technology, we thought of using a NAM microphone to capture sound from the area under the temporal bone (under the ear) without using traditional microphones that must be fixed in some way jointly with the user. Both of these solutions have been replaced by other more concrete and practical ones during the development and testing phases. The abandonment of neural network techniques is classically mainly attributable to the lack of data and their difficult production, or to the long-expected processing times.

The replacement of NAM, on the other hand, was caused by a negative evaluation by volunteer speakers suffering from various pathologies. This paragraph summarizes the attempts that have been made in order to trace the path taken in the direction of a performing algorithm applicable in real contexts.

7.3.1 Corpora and Audio Recording Modality

In this paragraph we describe the corpora built and exploited to perform tests with ML techniques. Indeed, the first crucial step consisted in the creation or finding of corpora suitable for training the networks. Before the abandonment of the NAM microphone, the networks would have been NAM-to-Whisper (NTW) and Whisper-to-Speech (WTS) transformations.

Then, the crucial point in training the Neural Networks (NNs) was to find or create NAM, normal, and whispered speech recordings aligned with each other for a wide range of speakers in order to train the NNs. A dataset composed of whispered and normal sentences (spontaneous, read, and elicited speech), from 11 able-bodied Northern-Italian speakers (5 males and 6 females) aged between 20 and 60 years, was used for preliminary analysis.

We created two different scripts: the first consisted in a list of 100 words, only read by 3 out of 11 speakers. The second was a sequence of 79 pairs of sentences, read by the remaining 8 speakers (for a total of 1264 sentences). Each pair was composed of a meaningful sentence and a nonsense phrase, expressly chosen to contain specific combinations of sounds and phonemes, covering the phonotactic possibilities of the Italian language (Cenceschi et al., 2021; Prieto et al., 2010–2014). The spontaneous speech has been recorded for seven speakers and consists in the description of a room or a favorite dish.

Recordings were made with a Steinberg UR22mkII sound card coupled to a tailor-made double microphone (comprising a dynamic and a NAM microphone realized by means of the stethoscope, see Figure 7.1).

FIGURE 7.1   A speaker wearing the tailor-made double microphone (NAM + dynamic capsule).

The dynamic headband microphone was located in front of the speaker's mouth, while the NAM was under the ear, in contact with the temporal bone. Then, the obtained stereo recording was split into mono to separate the NAM and the dynamic microphone channels. The resulting corpora are described below:

1. A corpus comprising normal and whispered speech, obtained through the dynamic microphone, and manually aligned.

2. A corpus comprising aligned sentences of *normal* speech recorded by means of the NAM and the dynamic microphone at the same time while the user is talking.

3. A corpus comprising aligned sentences of *whispered* speech recorded by means of the NAM and the dynamic microphone on a stereo at the same time while the user is talking.

4. *Fake* whisper speech samples obtained via Praat (Boersma, 2001) and processed via Praat Vocal Toolkit (Corretge, 2012–2021), exploiting the *To Whisper* feature. This passage was repeated both one and two times, to simulate different fake whispered speech.

In addition, 10 audiobook chapters narrated by 10 different speakers were downloaded from *LiberLiber*[9] and processed with the same methodology described in list item 4. All the sentences in these corpora have been manually split into syllables (listening and looking at the spectrogram) by means of Praat scripts for processing multiple files in TextGrid. Then, the obtained pairs of normal-whispered, NAM-whispered, and NAM-normal speech syllables have been aligned through Dynamic Time Warping, exploited in order to train different HUM configurations, and catalogued for future use.

Though the attention has been focused on healthy people corpora so far, the same recordings were also realized involving a speaker using esophageal speech (completely without vocal cord, see Meluzzi et al., 2022), and two patients affected by quadriplegia, suffering from difficulty in breathing due to no or limited mobility from the neck down. These samples have been used to test HUM on a set of various configurations and define the final user's profile. The following paragraph synthetizes the main NN architectures tested according to the typology of the input vocal signal to be enhanced.

7.3.2 NAM Microphone Signal and NNs

We initially thought to acquire the whispered speech on a NAM microphone, and then apply the two next steps. The first step resulted in a NAM-to-Whisper (NTW)[10] transformation followed by a Whisper-to-Speech (WTS)[11] passage, by applying a whisper enhancement method such as in Huang et al. (2019) and Zhou et al. (2012). Previous works suggest applying speech enhancement techniques (Malathi et al., 2019; Pascual et al., 2019b) to improve the intelligibility of the speech signal output.

However, despite literature review that seems to suggest promising results using small corpora (Parmar et al., 2019), the NAM did not prove to be usable both from a practical and a technical point of view. Firstly, all the feedback by people with various types of motor disabilities highlighted the need to not put anything on the speaker's neck. Moreover, the spectrum of the whispered speech signal obtained by means of the NAM is constrained below 3500 Hz, making any adequate whisper reconstruction difficult, such as obtaining data suitable for training the network.

After different attempts, the choice fell on the Minimum Mean Square Error Generative Adversarial Network (MMSE-GAN) and the PyTorch ML framework. We tried to retrieve the whisper spectrum from NAM signals basing both on MFCC (Mel filter bank for MFCC-FB40), and Magnitude Spectrum (Griffin-Lim phase) reconstruction. We considered both male plus female speakers and separate groups for different training sets. The method based on Magnitude Spectrum performed slightly better than the one based on MFCC, and spectrograms obtained from NAM samples showed some enhancement signs, but not enough robustness to be applied in real context.

7.3.3 Dynamic Microphone Signal and NNs

We abandoned the corpus obtained with a NAM microphone, and the technical choice fell on traditional dynamic microphone capsules, reducing the steps to a single WTS transformation based on MMSE-GAN. At this stage, we therefore exploited the data described in the paragraph dedicated to corpora also trying the use of the cited fake-whisper corpora.

However, the fake-whispered speech did not provide any added value, and we avoid deepening the description of these tests. Therefore, all the following architectures exploit the real whispered and normal-speech corpora (aligned syllables) mentioned above, including recited, elicited, and spontaneous speech. Regarding MFCC and Magnitude Spectrum reconstructions, both gave unsatisfactory results, although trying to reconstruct the Magnitude Spectrum seems to work better than MFCC, despite the higher dimension of the system. F0 prediction from a whisper tends to generalize better with small datasets, but the accuracy reached wasn't good enough to choose this solution.

Despite the fact that the WTS process provided spectrograms slightly better than those of the NAM-to-Whisper NN, the ML approach turned out to be unsatisfactory. In particular, generalizing the speech reconstruction process and making it work robustly for any speaker require an excessive amount of data in order to ensure a constant reliable reconstruction of the input.

7.4 GENERATIVE APPROACH DESIGN

In light of the previous results, we decided to completely overturn the approach and use the generative one, based on a tailor-made synthetic speech production algorithm. Results will be refined in the next months, but, to date, the resulting enhanced speech has come to an intelligibility slightly lower than that of a natural one. Further improvements will concern the quality of the timbre because its results remain unnatural, although it assumes acoustic characteristics that perceptually link it to that of the single speaker. The current resulting HUM flowchart is shown in Figure 7.2 and described below.

7.4.1 Implementation

The architecture of the system in its last version is defined as follows. The work has been developed in Python and rewritten and optimized in C++ programming language in a second step. First, the microphone signal is acquired at chunks of 1024 samples, 50% hop size, 44100 Hz sample rate. Each chunk is then passed through an anti-Larsen filter to avoid audio feedback, in order to remove the need for earphones and allow use through audio speakers. The anti-Larsen filter is a spectral filter that compares the mean of each magnitude band of the signal with its total Magnitude Spectrum; if the ratio of the power of a single band exceeds a threshold, a band-stop filter is set at the center frequency of the band, with a gain of −60 dB, to suppress the feedback.

The chunk is further filtered with a first-order high-pass filter (100 Hz cutoff frequency) and then split into overlapping frames of 256 samples each with 50% overlap. The frames are windowed with two different methods:

- Bartlett window for further audio usage of the data.

- Hamming window for Fast Fourier Transform (FFT) and Linear Predictive Coding (LPC) analysis.

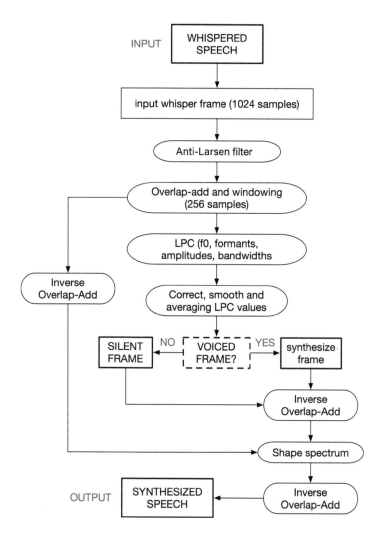

FIGURE 7.2   HUM flowchart.

The windowed frames are then stored into stacks along with the pre-viously computed frames, so that a 1024 sample chunk can always be reconstructed with an inverse overlap–add technique. Each new frame is further analyzed: the Root Mean Square (RMS) is computed, and formant frequencies, bandwidths, and amplitudes are calculated through LPC analysis.

Formant frequencies F1 and F2 are lowered by a fixed factor of 100 Hz for F1 and 150 Hz for F2, while F3 and F4 are left unaltered. Because of the instability of LPC with whispered speech, the resulting formant frequencies are rounded into the canonical approximation of frequency ranges of each formant (Kent & Vorperian, 2018), and these features are stacked into their respective data structures. The next step of the algorithm is to smooth the values of the features acquired with moving average filters, since there is too much variability in the neighboring frame features. F0 is then computed from the filtered RMS value according to methodology described in Morris & Clements (2002).

If a frame is considered to contain whispered speech, a corresponding synthesized frame is generated according to the feature values of the frame. The synthesizer uses the technique of additive synthesis to generate up to 80 partials, and the volume of each is set according to the formants of the frame.

A filtered noise (4000 to 20000 Hz) is then added to the sine mix, and the result is windowed with a Bartlett window. With corresponding whispered and synthesized, 3 overlapping frames (50% overlap) of 1024 samples are reconstructed with inverse overlap–add from the data stacks. This step is mandatory for the next processing because operating with 256 samples would generate too much variability in the resulting speech. The FFT is computed on the whispered frames, and two magnitude masks are calculated for each by averaging the Magnitude Spectrum with two different mean sizes, and then normalizing them. Each synthesized frame is transformed by FFT, each bin is multiplied by the two corresponding bins of the previous masks, then it is translated back to the time domain by Inverse FFT. The synthesized frames are finally windowed with a Bartlett window, and the output frame is calculated by inverse overlap–add.

7.5 RESULTS AND FUTURE WORK

Results were evaluated through a preliminary test of ten naïve listeners, and a large-scale evaluation will be performed according to robust perceptual test models. Currently, the output results were intelligible, almost in real time, characterized by a "full" timbre similar to normal speech mood. The aspect to be improved remains the timbre quality in terms of agreeableness and similarity to natural speech. The more the spoken input is delicate, the clearer the reconstructed speech becomes: the speaker's effort is thus minimized. Consequently, the user experience encourages the speaker not to emit too much air and minimize the uttering effort.

Therefore, the next step will be collecting subjective perceptive assessments through structured and validated tests. In order to base the evaluation on already tested methodologies, the state of the art was investigated regarding intelligibility and pleasantness of the reconstructed speech in different contexts (e.g., Finizia et al., 1998; Lagerberg et al., 2014;

Van Nuffelen et al., 2009; Walshe et al., 2008). In parallel a spectral tuning will be carried out based on the real differences between whispered and normal speech, to shape formants, bandwidths, and other spectral features to the characteristic values or variability thresholds of real speech.

To date, the objective and clinical identification of the target users who will benefit from HUM is under drafting. By addressing the solved problem, we can effectively identify various medical issues that give rise to it, allowing us to profile potential beneficiaries and anticipate their future needs in real-life situations. A final testing in real contexts will permit to set this target and move towards the commercial phase of the project.

HUM will then be injected in a miniaturized computer and VocalHUM's second step will begin. It will be focused on the design and construction of a tailor-made object. Some of the prerequisites have been collected during the first part of the project. Thus, a series of obsolete technological choices or poorly conforming to the future wearability of VocalHUM have already been excluded, so tracing the boundaries within which to move for the choice of components. For example, with regard to microphones, it will be necessary to place a miniaturized dynamic cardioid on the side of the mouth with a wire connection. It will be supported by an ultralight customizable headband, which depending on the position of the user can be replaced by a lateral single-ear support only. Regarding the micro-computer, the algorithm has been optimized for Raspberry Pi4, but a BELA[12] (audio-specific) will be tested soon.

Future research developments on the algorithmic side will expand the possible users to laryngectomized speakers. Until a few years ago, due to the lack of technology, laryngophones and throat microphones (Liua, H., & Ng, M. L. 2007) were only used to amplify the signal uttered by the patient (Cohen et al., 1984), without considering the possibility to employ these sources to perform spectral analysis and reconstruction. Laryngophones and throat microphones were then successfully applied to voice activity detection and speaker recognition (Sahidullah et al., 2016, 2017) and, if combined with acoustic microphones, they were also used to partially reconstruct the speech spectrum (Erzin, 2009; Shahina & Yegnanarayana, 2007; Turan, 2018; Zheng et al., 2003;). Despite extensive studies on this topic, the degraded speech produced by these patients is still very hard to understand. Although being psychologically important for users to have an acoustically acceptable voice, marketed tools are not so good from a qualitative and comfort point of view, at least not as much to reach large-scale success among patients. Thus, users are still constrained to employ invasive devices, since hardware products are still limited to old-fashioned tools like artificial larynx and simple signal amplifiers, and companies do not seem to be interested in implementing new solutions.

7.6 DISCUSSION AND CONCLUSION

HUM has proven to be fine for target users who normally articulate their speech but who

- Suffer from aphonia due to different, more or less severe and temporary causes such as pneumonia, severe reflux, and deterioration or partial removal of the vocal cords;
- Suffer from partial paralysis of the internal muscles and as a result are extremely tired when speaking.

Testing in real context could even extend this pool to laryngectomized speakers or ones who do not pronounce some phonemes but who need help to express themselves through a preestablished vocabulary (because they are not able to pronounce whole sentences). The generative approach permits us to apply a WTS enhancement in real time, which is less plausible through ML techniques. To date, results have been satisfactory in terms of voice quality and feasibility allowing to move forward from an applied research perspective.

Moreover, not only can HUM have immediate positive repercussions in clinics, but also in countless different areas whenever it is necessary to decode a muttered voice in real time, such as in pathology, forensics (e.g., in adverse investigative conditions), security (e.g., improving the speech of airplane pilots flying in critical situations), daily life, video game market, art, etc. The real-time algorithm for speech intelligibility enhancement will lay the groundwork for the development of further real-time applications. In particular, being for the first time integrated into a commercialized healthcare system will make the testing process easier on a large scale, overcoming the typical issue affecting many prototypes.

Moreover, HUM will be highly useful for collecting new whispered and murmured speech data to feed specific corpora, which are vital for medical research and ML technologies. Also, its algorithmic component will encourage speech processing research, laying the groundwork to countless real-time applications in different fields (e.g., security, IoT, AI, clinical, etc.).

VocalHUM will further lay the groundwork for real-time articulacy enhancement applications, specifically designed for aphonic and total laryngectomized speakers who do not wish to install invasive voice prostheses or use a laryngophone. Since laryngectomized users' speech presents different spectral information than the whispered/murmured speech, the algorithms developed during the first project will be upgraded and integrated to optimize the reconstruction, and in the management of unvoiced noises produced by the speaking apparatus.

NOTES

1. Rakshit, S. K., Keen, M. G., Bostick, J. E., & Ganci Jr, J. M. (2020). U.S. Patent No. 10,529,355. Washington, DC: U.S. Patent and Trademark Office; Kates, J. M. (1984). U.S. Patent No. 4,454,609; Zhang, M., Cao, K., Zeng, X., & Sun, H. (2020). U.S. Patent No. 10,811,030; Eisner, M., Huang, Z., & Duehren, D. (2015). U.S. Patent No. 14/561,026.
2. https://www.cerence.com
3. http://www.alango.com
4. https://advancedbionics.com/it/it/home.html
5. https://www.luminaud.com
6. http://www.griffinlab.com/catalog/
7. https://www.atosmedical.com
8. https://ultravoice.com/electrolarynx-speech-device-works/
9. https://www.liberliber.it/
10. NAM-to-Whisper is the process that transforms the signal collected by the NAM microphone to the corresponding whispered speech audio signal.
11. Whisper-to-Speech is the process that transforms whispered speech into more understandable speech.
12. https://bela.io/

REFERENCES

Amano-Kusumoto, A., & Hosom, J. P. (2011). A review of research on speech intelligibility and correlations with acoustic features. Center for Spoken Language Understanding, Oregon Health and Science University (Technical Report CSLU-011-001).

Berkovitz, R. (1982). Digital equalization of audio signals. In *Audio Engineering Society Conference: 1st International Conference: Digital Audio. Audio Engineering Society*, New York.

Boersma, P. (2001). Praat, a system for doing phonetics by computer. *Glot International, 5*(9), 341–345.

Bradlow, A. R., Torretta, G. M., & Pisoni, D. B. (1996). Intelligibility of normal speech I: Global and fine-grained acoustic-phonetic talker characteristics. *Speech Communication, 20*(3), 255.

Broeders, T. (2001, July). Forensic speech and audio analysis Forensic Linguistics 1998-2001. In *Proceedings 13th INTERPOL Forensic Science Symposium, Lyon, France D* (Vol. 2, pp. 54–84).

Cenceschi, S., Sbattella, L., & Tedesco, R. (2021). CALLIOPE: A multi-dimensional model for the prosodic characterization of information units. In *Estudios de Fonética Experimental, Journal of Experimental Phonetics*, E studios de phonética experimental (pp. 227–245).

Cohen, A., Van den Broecke, M. P., & Van Geel, R. C. (1984). A Study of Pitch Phenomena and Applications in Electrolarynx Speech. In *Speech and language* (Vol. 11, pp. 197–248). Elsevier, Amsterdam.

Corretge, R. (2012–2021). Praat vocal toolkit. http://www.praatvocaltoolkit.com

Doi, H., Nakamura, K., Toda, T., Saruwatari, H., & Shikano, K. (2010). Esophageal speech enhancement based on statistical voice conversion with Gaussian mixture models. *IEICE Transactions on Information and Systems, 93*(9), 2472–2482.

Erzin, E. (2009). Improving throat microphone speech recognition by joint analysis of throat and acoustic microphone recordings. *IEEE Transactions on Audio, Speech, and Language Processing, 17*(7), 1316–1324.

Finizia, C., Lindström, J., & Dotevall, H. (1998). Intelligibility and perceptual ratings after treatment for laryngeal cancer: Laryngectomy versus radiotherapy. *The Laryngoscope, 108*(1), 138–143.

French, N. R., & Steinberg, J. C. (1947). Factors governing the intelligibility of speech sounds. *The Journal of the Acoustical Society of America, 19*(1), 90–119.

Gao, M. (2002). Tones in whispered Chinese: Articulatory features and perceptual cues. (PhD Thesis, University of Victoria).

Godsill, S., Rayner, P., & Cappé, O. (2002). Digital Audio Restoration. In *Applications of Digital Signal Processing to Audio and Acoustics* (pp. 133–194). Springer, Boston, MA.

Huang, Y., Lian, H., Zhou, J., Wang, H., & Tao, L. (2019, September). An end to end method of whisper enhancement. In *2019 IEEE 2nd International Conference on Information Communication and Signal Processing (ICICSP)* (pp. 246–250), Weihai. IEEE.

Jovičić, S. T. (1998). Formant feature differences between whispered and voiced sustained vowels. *Acta Acustica united with Acustica, 84*(4), 739–743.

Kallail, K. J., & Emanuel, F. W. (1984). An acoustic comparison of isolated whispered and phonated vowel samples produced by adult male subjects. *Journal of Phonetics, 12*(2), 175–186.

Kent, R. D., & Vorperian, H. K. (2018). Static measurements of vowel formant frequencies and bandwidths: A review. *Journal of Communication Disorders, 74*, 74–97.

Lagerberg, T. B., Åsberg, J., Hartelius, L., & Persson, C. (2014). Assessment of intelligibility using children's spontaneous speech: Methodological aspects. *International Journal of Language & Communication Disorders, 49*(2), 228–239.

Lim, B. P. (2011). Computational differences between whispered and non-whispered speech. University of Illinois at Urbana-Champaign.

Liua, H., & Ng, M. L. (2007). Electrolarynx in voice rehabilitation. *Auris Nasus Larynx, 34*, 327–332.

Loizou, P. C., & Kim, G. (2010). Reasons why current speech-enhancement algorithms do not improve speech intelligibility and suggested solutions. *IEEE Transactions on Audio, Speech, and Language Processing, 19*(1), 47–56.

Malathi, P., Suresh, G. R., Moorthi, M., & Shanker, N. R. (2019). Speech enhancement via smart larynx of variable frequency for laryngectomee patient for Tamil language syllables using RADWT algorithm. *Circuits, Systems, and Signal Processing, 38*(9), 4202–4228.

Meluzzi, C., Cenceschi, S., Dani, F. R., & Trivilini, A. (2022). Phonetic characteristics of spontaneous speech in a total laryngectomized Italian speaker: Perspectives for speech enhancement algorithms. *In* Estudios de Fonética Experimental, Journal of Experimental Phonetics, (31), 45–58.

Morris, R. W., & Clements, M. A. (2002). Reconstruction of speech from whispers. *Medical Engineering & Physics, 24*(7–8), 515–520.

Olsson, J. (2018). *More Wordcrime: Solving Crime with Linguistics*. Bloomsbury Publishing.

Parmar, M., Doshi, S., Shah, N. J., Patel, M., & Patil, H. A. (2019, September). Effectiveness of cross-domain architectures for whisper-to-normal speech conversion. In *2019 27th European Signal Processing Conference (EUSIPCO)* (pp. 1–5), Coruña. IEEE.

Pascual, S., Serrà, J., & Bonafonte, A. (2019a). Towards generalized speech enhancement with generative adversarial networks. *arXiv preprint arXiv:1904.03418*.

Pascual, S., Serra, J., & Bonafonte, A. (2019b). Time-domain speech enhancement using generative adversarial networks. *Speech Communication, 114*, 10–21.

Patel, M., Parmar, M., Doshi, S., Shah, N., & Patil, H. A. (2019). Novel inception-GAN for whisper-to-normal speech conversion.

Prieto, P., Borràs-Comes, J., & Roseano, P. (Coords.) (2010–2014). Interactive atlas of romance intonation. Web page: <http://prosodia.upf.edu/iari/>.

Riekhakaynen, E. I. (2020, January). Corpora of Russian spontaneous speech as a tool for modelling natural speech production and recognition. In *2020 10th Annual Computing and Communication Workshop and Conference (CCWC)* (pp. 0406–0411), Las Vegas. IEEE.

Robinson, T., Hochberg, M., & Renals, S. (1994, April). IPA: Improved phone modelling with recurrent neural networks. In *Proceedings of ICASSP'94. IEEE International Conference on Acoustics, Speech and Signal Processing* (Vol. 1, pp. I–37), Adelaide. IEEE.

Sahidullah, M., Gonzalez Hautamäki, R., Lehmann, T. D. A., Kinnunen, T., Tan, Z. H., Hautamäki, V., & Pitkänen, M. (2016). Robust speaker recognition with combined use of acoustic and throat microphone speech.

Sahidullah, M., Thomsen, D., Hautamäki, L., Kinnunen, R. G., Tan, T., Parts, Z. H., & Pitkänen, R. (2017). Robust voice liveness detection and speaker verification using throat microphones. *IEEE/ACM Transactions on Audio, Speech, and Language Processing, 26*(1), 44–56.

Shah, N. J., & Patil, H. A. (2020). 5 non-audible murmur to audible speech conversion. *Voice Technologies for Speech Reconstruction and Enhancement, 6*, 125.

Shahina, A., & Yegnanarayana, B. (2007). Mapping speech spectra from throat microphone to close-speaking microphone: A neural network approach. *EURASIP Journal on Advances in Signal Processing, 2007, (1)*, 1–10.

Sharifzadeh, H. R., McLoughlin, I. V., & Russell, M. J. (2010). Toward a comprehensive vowel space for whispered speech. In *2010 7th International Symposium on Chinese Spoken Language Processing*, (pp. 65–68), Taiwan. IEEE.

Strassel, S., Miller, D., Walker, K., & Cieri, C. (2003). Shared resources for robust speech-to-text technology. In Eighth European Conference on Speech Communication and Technology, Geneva.

Stylianou, Y. (1996). Harmonic plus noise models for speech, combined with statistical methods, for speech and speaker modification. PhD thesis, Ecole Nationale Superieure des Telecommunications.

Sun, N., Wei, L., Wang, H., Wang, X., Gao, M., Hu, X., & Shi, S. (2021). Qualitative study of the psychological experience of COVID-19 patients during hospitalization. *Journal of Affective Disorders*, 278, 15–22.

Toda, T., Nakagiri, M., & Shikano, K. (2012). Statistical voice conversion techniques for body-conducted unvoiced speech enhancement. *IEEE Transactions on Audio, Speech, and Language Processing*, 20(9), 2505–2517.

Toda, T., & Shikano, K. (2005). NAM-to-speech conversion with Gaussian mixture models, Proc. Interspeech 2005, (pp. 1957–1960), Lisbon.

Tran, V. A., Bailly, G., Loevenbruck, H., & Toda, T. (2009, September). Multimodal HMM-based NAM-to-speech conversion. In *Interspeech 2009–10th Annual Conference of the International Speech Communication Association* (pp. 656–659), Brtighton.

Tsunoda, K., Ohta, Y., Niimi, S., Soda, Y., & Hirose, H. (1997). Laryngeal adjustment in whispering: Magnetic resonance imaging study. *Annals of Otology, Rhinology & Laryngology*, 106(1), 41–43.

Turan, M. A. T. (2018). Enhancement of throat microphone recordings using Gaussian mixture model probabilistic estimator. *arXiv preprint arXiv:1804.05937.*

Van Nuffelen, G., Middag, C., De Bodt, M., & Martens, J. P. (2009). Speech technology-based assessment of phoneme intelligibility in dysarthria. *International Journal of Language & Communication Disorders*, 44(5), 716–730.

Walshe, M., Miller, N., Leahy, M., & Murray, A. (2008). Intelligibility of dysarthric speech: Perceptions of speakers and listeners. *International Journal of Language & Communication Disorders*, 43(6), 633–648.

Weitzman, R. S., Sawashima, M., Hirose, H., & Ushijima, T. (1976). Devoiced and whispered vowels in Japanese. *Annual Bulletin, Research Institute of Logopedics and Phoniatrics*, 10(61–79), 29–31.

Zheng, Y., Liu, Z., Zhang, Z., Sinclair, M., Droppo, J., Deng, L., & Huang, X. (2003, November). Air-and bone-conductive integrated microphones for robust speech detection and enhancement. In *2003 IEEE Workshop on Automatic Speech Recognition and Understanding (IEEE Cat. No. 03EX721)* (pp. 249–254), St Thomas. IEEE.

Zhou, J., Liang, R., Zhao, L., & Zou, C. (2012). Whisper intelligibility enhancement using a supervised learning approach. *Circuits, Systems, and Signal Processing*, 31(6), 2061–2074.

CHAPTER 8

Analysis of Features for Machine Learning Approaches to Parkinson's Disease Detection

Claudio Ferrante, Licia Sbattella, and Vincenzo Scotti

DEIB, Politecnico di Milano, Milan, Italy

Bindu Menon

Apollo Specialty Hospitals, Nellore, Andhra Pradesh, India

Anitha S. Pillai

Hindustan Institute of Technology and Science, Kelambakkam, Tamil Nadu, India

8.1 INTRODUCTION

Deep Learning (DL) has slowly become an essential tool for Natural Language Processing (NLP). Apart from the impressive results on text processing [1,2], DL models allowed to improve the results of multiple speech-related applications, like Automatic Speech Recognition (ASR) [3,4], speaker identification [5], conditioned Text-to-Speech (TTS) synthesis [6–8] or speech emotion recognition [9].

DOI: 10.1201/9781003296126-13

These DL models are particularly useful in the presence of small datasets or under-resourced problems (in terms of data and domain-knowledge availability). They allow to exploit techniques like transfer learning and fine-tuning [10], where the features computed by a DL neural network model trained on a large generic dataset are re-used on a specific problem with a smaller dataset, generally resulting in improved performances.

In this work, we focus on Parkinson's disease detection from speech. We propose a probabilistic classification pipeline to detect if a patient is affected by Parkinson's disease by analyzing voice recordings. To evaluate the generalization capabilities enabled by DL models for audio/speech feature extraction, we test them on a relatively small dataset of samples in Telugu. These settings represent a challenge due to the reduced dataset size and the scarce availability of language-specific analysis models.

We divide this chapter into the following sections. In Section 8.2 we present the related works in terms of features for speech analysis and results on Parkinson's disease detection from voice. In Section 8.3 we present the abstract classification pipeline we adopted, suggesting possible implementations of the various modules. In Section 8.4 we describe the dataset we collected to train and evaluate different classification models. In Section 8.5 we describe the implemented pipeline configurations we evaluated, and we report the results obtained during the evaluation. Finally, in Section 8.6 we summarize our work and suggest possible future evolution.

8.2 RELATED WORKS

In this section, we present the features commonly employed for speech classification, both traditional and DL based.

Additionally, we present the latest results for Parkinson's disease detection from speech.

8.2.1 Features for Speech Analysis

Traditionally, the speech features adopted for NLP are divided into two groups: prosodic and acoustic features [11,12]. The former group includes features used to describe peculiarities of speech, such as: Pitch, Intensity, Harmonicity, Jitter, Shimmer, Speech Rate, Short-Term Energy, Short-Term Entropy, etc. The latter group includes features used to describe the acoustic properties of speech, such as: Spectrogram (magnitude or power), Mel-spectrogram, Mel Frequency Cepstral Coefficients (MFCC), Spectrogram statistics (centroid, spread, flux, rolloff, entropy), Chromagram, etc.

More recent approaches, instead, propose to re-use DL models trained on large data collections. The internal representations learned by these models are particularly informative and can be directly transferred or easily adapted to many new problems. The most popular models in this sense are SoundNet [13], VGGish [14], and Wav2Vec [3, 15]. The first two are very generic models, though for acoustic analysis not necessarily aimed at speech. SoundNet is a 1D Convolutional Neural Network (CNN) trained to predict from the audio track of video clips the pseudo-labels generated from an object recognition deep neural network and a scene recognition deep neural network that processed the images of the video clips. VGGish, on the other hand, is a 2D CNN trained on a large audio classification dataset containing a large number of labels and samples. Instead, both versions of Wav2Vec were specifically designed for speech analysis problems and were originally used as input for state-of-the-art ASR models.

8.2.2 Parkinson's Disease Detection from Speech

Parkinson's disease detection from speech has already been explored as a Machine Learning (ML) problem.

More sophisticated solutions also adopted dimensionality reduction techniques to feed more compact and informative feature vectors (encoding the input speech signal) to the classifiers. Different classification algorithms, like Artificial Neural Networks, Support Vector Machines, and k-Nearest Neighbors, have been adopted for this detection problem [16–23]. Usually, these solutions involved the extraction of prosodic and acoustic features (mainly MFCC, Jitter, Shimmer, and Pitch), which allowed to train discriminative models with impressive results. In some cases, these features were further processed through dimensionality reduction transformations to keep only the most relevant components of the vectors encoding the audio clips to classify, which resulted in further improvements is some cases.

Recent results also explored the effect of DL features to train a classifier for this Parkinson's disease detection problem [19], reaching more than 80% recognition accuracy on a dataset of English audio clips, outperforming the other considered classifiers based on spectral features. Other works started focusing on building classifiers compatible with multiple languages [20]; the results showed that acoustic and spectral features can be used to build high-performing classifiers (reaching more than 90% recognition accuracy) on English and Italian.

All these works relied on larger datasets like Mobile Device Voice Recordings at King's College London (MDVR-KCL) from both early and advanced Parkinson's disease patients and healthy controls [24] and Italian Parkinson's Voice and Speech [25,26], which account for more than 1 h of recordings. In our case, we are dealing with a much smaller dataset; thus we are interested in seeing if and how much performance degrades when using similar classification pipelines.

8.3 PROPOSED MACHINE LEARNING PIPELINE

In this section, we describe the classification pipeline we propose for Parkinson's disease detection from speech; we depicted the pipeline in Figure 8.1. We compose each of the stages of the pipeline (pre-processing, feature extraction, and classification) of different modules. The choice of specific module implementations allows to instance the proposed pipeline into a classification model, which can be trained and evaluated.

8.3.1 Pre-Processing

In our pipeline, we considered two pre-processing steps: segmentation and denoising. Both steps prepare the raw data for the feature extraction stage.

Segmentation consists of the splitting of the audio clip in presence of longer pauses, which generally mark the end of an utterance. This step can be done by hand (however, it may require a lot of time) or automatically. In the latter case, it is better to do it after denoising, to avoid errors due to additional sounds present in the recording that overlap with the voice.

The denoising module takes care of removing (as much as possible) additional signals in the input audio clip which overlap with the voice to analyze. State-of-the-art solutions use DL models trained on many hours of

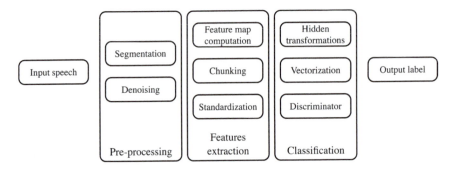

FIGURE 8.1 Visualization of the abstract classification pipeline.

data and can achieve impressive results. This is an important step; due to the reduced size of the dataset we are working with, we cannot expect the classification models to generalize and learn implicitly to ignore the noise.

8.3.2 Features Extraction

Features extraction is the core stage of our pipeline. The main step at this stage is the computation of the feature map, which, after the appropriate post-processing steps is the actual input of the classification model.

Each pre-processed input speech signal undergoes a transformation to extract a feature map. Traditionally data samples for ML are represented by a $d \in \mathbb{N}$ feature vector. However, for some problems like speech or image analysis, it is possible to leverage the spatial structure of the input and generate a feature map. In our case, from a speech signal, we can compute a feature map that is a sequence of feature vectors (each computed in a specific time window of the input signal) that can be encoded in a matrix $X \in \mathbb{R}^{t \times d}$, where t is the number of time positions and d is the number of features for each vector. DL models, as well as algorithms to compute traditional prosodic and acoustic features, iteratively transform the raw input signal to obtain a feature map.

Depending on the duration of the input audio clip, we considered the possibility of an intermediate chunking module. To avoid processing too large segments of audio, which can be computationally expensive and may harm the results in the presence of shorter input, we introduced an optional chunking step. The feature maps can be chunked into smaller windows along the time axis, to process smaller portions of the input speech. In this way, from the same segment, it is possible to extract multiple samples for the classifier.

The last step in the features extraction stage is standardization. This transformation is used to mean-center the data and impose a variance of 1 for all the features individually. Standardization is used to have the same scale on all the features, which helps the convergence and stability of the learning algorithm used for classification. There exists some robust version of this transformation using, for example, the median instead of the mean for centering.

8.3.3 Classification

The last step of the proposed pipeline is a CNN classifier [27]. We approach the problem as a supervised binary classification problem. In input, we have a feature map extracted from an audio clip of human speech, and in output, we have the probability for that clip to correspond to a Parkinson's disease patient. We report a diagram of the considered architecture in Figure 8.2.

The CNN is composed of a hidden transformation $h(\cdot)$, composed of convolutional blocks. These transformations process the input feature map $\boldsymbol{X} \in \mathbb{R}^{t \times d}$ transforming it into another feature map $\boldsymbol{H} \in \mathbb{R}^{t \times h}$ to be used for the final classification. The convolutional blocks contain, in order, a 1D convolutional layer, a $ReLU(\cdot)$ activation and a max pooling

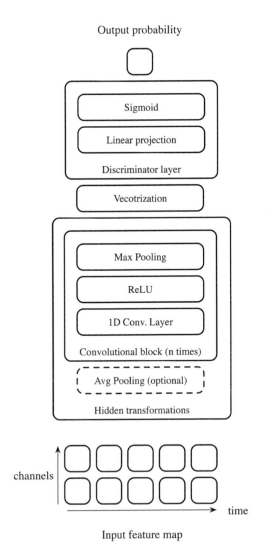

FIGURE 8.2   Convolutional Neural Network classifier architecture.

transformation to reduce the spatial dimensionality of the feature map. Optionally, we considered an initial average pooling transformation to reduce the spatial dimensionality before the convolutional blocks. This optional transformation is useful when dealing with highly dense feature maps like those from handcrafted features or from Wav2Vec 2.0.

After the hidden transformation, we apply a vectorization transformation. The role of this transformation is to drop the spatial dimension and convert the feature map $H \in \mathbb{R}^{t \times h}$ into a feature vector $h \in \mathbb{R}^{h}$ to be classified. We considered the following vectorization approaches: flattening (or unrolling), Global Average Pooling (GAP) [28], and Global Max Pooling (GMP) [28].

The last layer of the CNN is a linear transformation that maps the feature vector $h \in \mathbb{R}^{h}$ into a scalar value. This value is passed through a sigmoid activation function to have the output probability score.

8.4 DATA

As discussed, in this work we adopted a dataset of audio clips collected from Telugu speakers. Two reasons behind this choice are: we wanted to establish a baseline, and we wanted to see if DL features for speech analysis allow generalizing on under-resourced languages and small datasets, like in this case.

The dataset we adopted is composed of two parts. The samples from Parkinson's disease patients come from a private dataset, composed of 12 m 39 s of audio recordings. To balance this dataset with samples from healthy persons (in the sense of speakers not affected by Parkinson's disease), we gathered the audio samples from the delta segment of the Telugu split of the Open Speech and Language Resources (OpenSLR) corpus [29], which accounts for 15 m 35 s of recordings. The total amount of available data is 28 m 14 s. In terms of samples, we collected a total 281 audio clips, 200 from patients in healthy conditions and 71 from patients affected by Parkinson's disease.

Before processing the audio clips with the feature extraction modules of our pipeline, we manually segmented those clips. We split the recordings on longer pauses, which we associated with utterance boundaries. In Figure 8.3 we displayed the distribution of the audio clip lengths after the manual segmentation. As can be seen, audio clips of Parkinson's disease patients cover a wider duration range than those from healthy patients. This difference is primarily due to the different sources of the data samples, other than the articulation difficulties due to the disease.

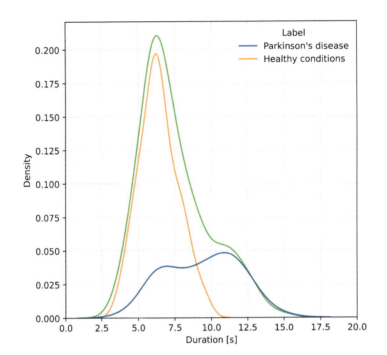

FIGURE 8.3 Distribution of audio clips duration after segmentation. The green line is the overall distribution including all samples.

As we explained in Section 8.3, the speech samples are analyzed in smaller chunks, to prevent overfitting of the classifiers, given the reduced size of the dataset.

8.5 EVALUATION AND RESULTS

In this section, we describe the experiments we conducted on different models, and how we evaluated their performances. Additionally, we present and comment on the obtained results.

8.5.1 Experiments

Our experiments were mainly to compare different modules for the feature extraction stage, in order to identify the most suitable features for Parkinson's disease from speech on the considered Telugu dataset. In this sense, we compared different feature map computation approaches, analyzing the results obtained using features from different DL models. Additionally, we explored different vectorization algorithms, applying them to the extracted feature maps.

In the pre-processing stage, after the manual segmentation step to isolate the different utterances in the same recording, we used a denoising application to enhance voice and remove background noises that could have affected the classifier. We employed the RNNoise [30] tool to denoise the input speech segments.

For the feature extraction stage, we compared three different pre-trained DL models for acoustic features extraction: SoundNet, VGGish, and Wav2Vec 2.0. To have a term of comparison with these DL features, we also trained some models using traditional (handcrafted) speech analysis features. Following recent work on Parkinson's disease detection from speech in English [20], we adopted the following prosodic and acoustic features: MFCC, Pitch, Jitter (absolute, relative, rap, and PPQ5), Shimmer (absolute, absolute dB, relative, APQ3, and APQ5), Harmonicity. We concatenated these prosodic and acoustic features into a single, d-dimensional, feature map.

We applied chunking to the feature maps resulting from features extraction using non-overlapping windows of 4 s. We applied padding to make sure that all windows ended up composed of 4 s data. We extended both sides of the feature map (before the chunking step), replicating the value on the border. We repeated the values so that the input sequence of features could be decomposed into an integer number of chunks. For all the considered input features, we applied standard scaling, computing mean and variance of the individual feature vectors composing all the feature maps.

After chunking, we obtained 538 samples, 315 from healthy people and 223 from patients affected by Parkison's disease. We applied minority oversampling to balance the dataset.

Concerning the CNN, we used a standard stack of 1D convolutional layers with non-linear activation as hidden transformation, followed by the vectorization operation and a final linear classification layer. As anticipated in Section 8.3.3, we considered flattening, GAP, and GMP as vectorization transformations. For each input feature-vectorization algorithm pair, we searched for the best hyper-parameters configuration of the CNN using five-fold cross-validation. We considered 1, 2, or 3 convolutional blocks and either 512 or 1024 output channels for the convolutions. We used a constant kernel width of 3 for all configurations. We trained the CNN using the Adam optimizer, in the cross-validation step we considered as learning rates 10^{-3} and $5 \cdot 10^{-4}$. To prevent overfitting, we used dropout with a probability of 10%.

TABLE 8.1   Confusion Matrix

	Predicted Positive	**Predicted Negative**
Labelled positive	*TP*	*FN*
Labelled negative	*FP*	*TN*

8.5.2 Evaluation Approach

To ensure a correct evaluation of the model performances, we split the audio segments into train and test, with a 75–25% split. We used the same training and testing subsets with each of the proposed models (i.e., pipeline implementations). For each model, we computed the common metrics used in ML for classification and information retrieval problems [31], defined starting from the confusion matrix.

Referring to the confusion matrix in Table 8.1, we introduce the following definitions:

- TP = True Positive (i.e., positive values correctly predicted as such)

- TN = True Negative (i.e., negative values correctly predicted as such)

- FP = False Positive (i.e., negative values predicted as positive)

- FN = False Negative (i.e., positive values predicted as negative)

Given that this is a Parkinson's disease detection problem, we associate the positive class with the disease condition and the negative class with the healthy condition.

To assess the quality of the trained classification models, we computed the following metrics:

- Accuracy $= \frac{TP+TN}{TP+TN+FP+FN}$

- Precision $= \frac{TP}{TP+FP}$ (i.e., positive predictive value)

- Recall $= \frac{TP}{TP+FN}$ (i.e., sensitivity, hit rate, true positive rate)

- Specificity $= \frac{TN}{TN+FP}$ (i.e., selectivity, negative class recall true negative rate)

- F_1-score $= \frac{2 \cdot \text{precision} \cdot \text{recall}}{\text{precision}+\text{recall}}$ (i.e., sensitivity, hit rate, true positive rate)

- AUC (Area Under the Curve of the Receiver Operating Characteristic)

8.5.3 Results and Comments

We reported the measured metrics on the test split in Figure 8.4. All models managed to achieve good performances despite the reduced dataset size: in most cases, we achieved scores for all metrics >95%. All features showed to be independent from the pooling approach, reaching similar results across the different vectorizations algorithms.

Concerning DL features, VGGish and Wav2Vec 2.0 achieve the overall best results. On the other hand, SoundNet produced the worst results. In all cases where the classifiers produce worse results, we can notice that the recall score is lower than the precision one. Thus, we can hypothesize that, in those cases, the unbalance in the number of negative samples (corresponding to healthy patients) influenced negatively the model, causing the increase of false negatives.

Interestingly, the handcrafted features perform comparably to VGGish and Wav2Vec 2.0. In fact, despite the small dataset, we managed to achieve

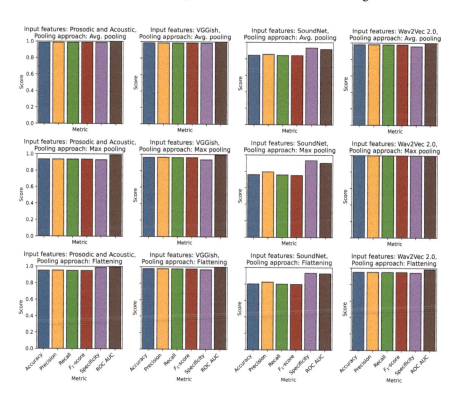

FIGURE 8.4   Results of the different tested configurations. Each row corresponds to a different pooling (vectorization) approach, each column corresponds to an input feature.

almost a perfect score (which is not possible). This hints that the few hand-crafted features encode very useful information for the task.

8.6 CONCLUSION

In this chapter, we approached a speech analysis problem, Parkinson's disease detection from voice, using a ML pipeline. We evaluated different feature extraction approaches, comparing traditional prosodic and acoustic features against features computed by deep neural networks. We used the extracted features to fit a CNN classifier. All the experiments were conducted on a relatively small dataset of audio clips collected from Telugu speakers. We achieved equally good results using both deep features and, surprisingly, using handcrafted features. The reported scores are in line with those achieved on bigger datasets, showing that even with low resources DL models can yield good generalization capabilities. Nevertheless, handcrafted features showed to be capable of yielding valid results, comparable to those of the deep models, despite the reduced dataset size. This hints that DL solutions for audio processing still need to become an irreplaceable tool. For the sake of reproducibility, we are sharing the source code via *GitHub*[1].

Concerning future direction, we are willing to explore different learning paradigms to improve the detection results. On one hand, we are considering exploiting regularities within data and seeing if unsupervised learning may lead to better results. Ideally, the similarities and dissimilarities between samples may be used to feed a clustering algorithm, allowing, possibly, to group automatically samples from healthy patients and patients affected by Parkinson's disease. On the other hand, we are considering anomaly detection approaches. In fact, given that there are many more available samples of speech from healthy people, it would be possible to detect samples of speech from Parkinson's disease patients as outliers to the distribution of the regular data.

NOTE

1. https://github.com/vincenzo-scotti/voice_analysis_parkinson

REFERENCES

1. C. Raffel, N. Shazeer, A. Roberts, K. Lee, S. Narang, M. Matena, Y. Zhou, W. Li and P. J. Liu, "Exploring the Limits of Transfer Learning With a Unified Text-to-Text Transformer," *J. Mach. Learn. Res*, vol. 21, pp. 140:1–140:67, 2020.

2. T. B. Brown, B. Mann, N. Ryder, M. Subbiah, J. Kaplan, P. Dhariwal, A. Neelakantan, P. Shyam, G. Sastry, A. Askell, S. Agarwal, A. Herbert-Voss, G. Krueger, T. Henighan, R. Child, A. Ramesh, D. M. Ziegler, J. Wu, C. Winter, C. Hesse, M. Chen, E. Sigler, M. Litwin, S. Gray, B. Chess, J. Clark, C. Berner, S. McCandlish, A. Radford, I. Sutskever and D. Amodei, "Language Models are Few-Shot Learners," in *Advances in Neural Information Processing Systems 33: Annual Conference on Neural Information Processing Systems 2020, NeurIPS 2020, December 6-12, 2020, virtual*, 2020.

3. A. Baevski, Y. Zhou, A. Mohamed and M. Auli, "wav2vec 2.0: A Framework for Self-Supervised Learning of Speech Representations," in *Advances in Neural Information Processing Systems 33: Annual Conference on Neural Information Processing Systems 2020, NeurIPS 2020, December 6–12, 2020, virtual*, 2020.

4. A. Radford, J. W. Kim, T. Xu, G. Brockman, C. McLeavey and I. Sutskever, "Robust Speech Recognition via Large-Scale Weak Supervision," *CoRR*, vol. abs/2212.04356, 2022, https://arxiv.org/abs/2212.04356.

5. L. Wan, Q. Wang, A. Papir and I. Lopez-Moreno, "Generalized End-to-End Loss for Speaker Verification," in *2018 IEEE International Conference on Acoustics, Speech and Signal Processing, ICASSP 2018, Calgary, AB, Canada, April 15–20, 2018*, 2018.

6. R. J. Skerry-Ryan, E. Battenberg, Y. Xiao, Y. Wang, D. Stanton, J. Shor, R. J. Weiss, R. Clark and R. A. Saurous, "Towards End-to-End Prosody Transfer for Expressive Speech Synthesis with Tacotron," in *Proceedings of the 35th International Conference on Machine Learning, ICML 2018, Stockholmsmässan, Stockholm, Sweden, July 10–15, 2018*, 2018.

7. A. Favaro, L. Sbattella, R. Tedesco and V. Scotti, "ITAcotron 2: Transfering English Speech Synthesis Architectures and Speech Features to Italian," in *Proceedings of The Fourth International Conference on Natural Language and Speech Processing (ICNLSP 2021)*, Trento, 2021.

8. A. Favaro, L. Sbattella, R. Tedesco and V. Scotti, "ITAcotron 2: The Power of Transfer Learning in Expressive TTS Synthesis," in M. Abbas (ed), *Analysis and Application of Natural Language and Speech Processing*, A cura di, Cham, Springer International Publishing, 2022, p. 1–20.

9. V. Scotti, F. Galati, L. Sbattella and R. Tedesco, "Combining Deep and Unsupervised Features for Multilingual Speech Emotion Recognition," in *Pattern Recognition. ICPR International Workshops and Challenges - Virtual Event, January 10–15, 2021, Proceedings, Part II*, 2020.

10. J. Yosinski, J. Clune, Y. Bengio e and H. Lipson, "How transferable are features in deep neural networks?," in *Advances in Neural Information Processing Systems 27: Annual Conference on Neural Information Processing Systems 2014, December 8–13 2014, Montreal, Quebec, Canada*, 2014.

11. T. Giannakopoulos, "pyAudioAnalysis: An Open-Source Python Library for Audio Signal Analysis," *PLOS ONE*, vol. 10, pp. 1–17, December 2015.

12. V. Chernykh, G. Sterling and P. Prihodko, "Emotion Recognition from Speech with Recurrent Neural Networks," *CoRR*, vol. abs/1701.08071, 2017, https://arxiv.org/abs/1701.08071

13. Y. Aytar, C. Vondrick and A. Torralba, "SoundNet: Learning Sound Representations from Unlabeled Video," in *Advances in Neural Information Processing Systems 29: Annual Conference on Neural Information Processing Systems 2016, December 5–10, 2016, Barcelona, Spain*, 2016.

14. S. Hershey, S. Chaudhuri, D. P. W. Ellis, J. F. Gemmeke, A. Jansen, R. C. Moore, M. Plakal, D. Platt, R. A. Saurous, B. Seybold, M. Slaney, R. J. Weiss and K. W. Wilson, "CNN architectures for large-scale audio classification," in *2017 IEEE International Conference on Acoustics, Speech and Signal Processing, ICASSP 2017, New Orleans, LA, USA, March 5–9, 2017*, 2017.

15. S. Schneider, A. Baevski, R. Collobert and M. Auli, "wav2vec: Unsupervised Pre-Training for Speech Recognition," in *Interspeech 2019, 20th Annual Conference of the International Speech Communication Association, Graz, Austria, 15–19 September 2019*, 2019.

16. J. R. Williamson, T. F. Quatieri, B. S. Helfer, J. Perricone, S. S. Ghosh, G. A. Ciccarelli and D. D. Mehta, "Segment-dependent dynamics in predicting parkinson's disease," in *INTERSPEECH 2015, 16th Annual Conference of the International Speech Communication Association, Dresden, Germany, September 6–10, 2015*, 2015.

17. B. Karan, S. S. Sahu and K. Mahto, "Parkinson Disease Prediction Using Intrinsic Mode Function Based Features from Speech Signal," *Biocybernetics and Biomedical Engineering*, vol. 40, pp. 249–264, 2020.

18. W. Rahman, S. Lee, M. S. Islam, V. N. Antony, H. Ratnu, M. R. Ali, A. Al Mamun, E. Wagner, S. Jensen-Roberts, E. Waddell and others, "Detecting Parkinson Disease Using a Web-Based Speech Task: Observational Study," *Journal of Medical Internet Research*, vol. 23, p. e26305, 2021.

19. S. Kurada and A. Kurada, "Poster: Vggish Embeddings Based Audio Classifiers to Improve Parkinson's Disease Diagnosis," in *5th IEEE/ACM International Conference on Connected Health: Applications, Systems and Engineering Technologies, CHASE 2020, Crystal City, VA, USA, December 16–18, 2020*, 2020.

20. A. A. Toye and S. Kompalli, "Comparative Study of Speech Analysis Methods to Predict Parkinson's Disease," *CoRR*, vol. abs/2111.10207, 2021, https://arxiv.org/abs/2111.10207.

21. A. Favaro, S. Motley, Q. M. Samus, A. Butala, N. Dehak, E. S. Oh and L. Moro-Velazquez, Artificial Intelligence Tools to Evaluate Language and Speech Patterns in Alzheimer's Disease, *Alzheimer's & Dementia*, vol. 18, no. S2, p. e064913, 2022.

22. A. Favaro, C. Montley, M. Iglesias, A. Butala, E. S. Oh, R. D. Stevens, J. Villalba, N. Dehak and L. Moro-Velasquez, "A Multi-Modal Array of Interpretable Features to Evaluate Language and Speech Patterns in Different Neurological Disorders," in IEEE Spoken Language Technology Workshop (SLT), Doha, Qatar, 2022.

23. G. Dimauro, V. D. Nicola, V. Bevilacqua, D. Caivano, and F. Girardi, "Assessment of Speech Intelligibility in Parkinson's Disease Using a Speech-To-Text System," *IEEE Access*, vol. 5, pp. 22199–22208, 2017.

24. H. Jaeger, D. Trivedi, and M. Stadtschnitzer, Mobile Device Voice Recordings at King's College London (MDVR-KCL) from both early and advanced Parkinson's disease patients and healthy controls, Zenodo, 2019.

25. G. Dimauro, and F. Girardi, Italian Parkinson's Voice and Speech, *IEEE Dataport*, 2019.

26. G. Dimauro, D. Caivano, V. Bevilacqua, F. Girardi and V. Napoletano, "VoxTester, software for digital evaluation of speech changes in Parkinson disease," in *2016 IEEE International Symposium on Medical Measurements and Applications, MeMeA 2016, Benevento, Italy, May 15–18, 2016*, 2016.

27. I. J. Goodfellow, Y. Bengio, and A. C. Courville, *Deep Learning*, MIT Press, Cambridge, MA, 2016.

28. M. Lin, Q. Chen, and S. Yan, "Network In Network," in *2nd International Conference on Learning Representations, ICLR 2014, Banff, AB, Canada, April 14–16, 2014, Conference Track Proceedings*, 2014.

29. F. He, S.-H. C. Chu, O. Kjartansson, C. Rivera, A. Katanova, A. Gutkin, I. Demirsahin, C. Johny, M. Jansche, S. Sarin, and K. Pipatsrisawat, "Opensource Multi-speaker Speech Corpora for Building Gujarati, Kannada, Malayalam, Marathi, Tamil and Telugu Speech Synthesis Systems," in *Proceedings of the Twelfth Language Resources and Evaluation Conference*, LREC 2020, Marseille, France, May 11-16, 2020.

30. J.-M. Valin, "A Hybrid DSP/Deep Learning Approach to Real-Time Full-Band Speech Enhancement," in 20th IEEE International Workshop on Multimedia Signal Processing, MMSP 2018, *Vancouver, BC, Canada*, August 29–31, 2018, 2018.

31. M. J. Zaki, and W. Meira, *Data Mining and Machine Learning: Fundamental Concepts and Algorithms*, 2nd Edition, Cambridge University Press, Cambridge (England), 2020.

CHAPTER 9

Conversational Agents, Natural Language Processing, and Machine Learning for Psychotherapy

Licia Sbattella

Politecnico di Milano, Milan, Italy

9.1 INTRODUCTION

Mental illness is one of the most pressing public health issues of our time. Economic constraints, social stigma, and scarce availability of professionals require, on one hand, to augment clinical support and quality, and on the other hand, to create instruments able to augment treatment and to enrich training and supervision methods.

Psychotherapy and its clinical interactions should also be considered a rich and special field for testing Natural Language Processing (NLP) and Machine Learning (ML) research efforts, with a special attention to the algorithms used to realize Conversational Agents (CAs) as adjoint therapists or supervisors, supporting different aspects of the therapeutic process (diagnosis, treatment, evaluation of the process and its results, support for training and supervision of specialists).

The use of CAs in the field of mental health and psychotherapy is in the early stages of development, particularly when compared with other application sectors. This chapter helps to understand the reasons of this

DOI: 10.1201/9781003296126-9

delay and suggests some strategies that could be adopted to compensate for it, paying particular attention to the quality of the psychotherapeutic process.

The chapter analyzes and discusses literature dealing with CAs supporting the mental health domain, with a special focus on psychotherapy and psychological support, evaluation of strategies and systems, analysis of clinical interactions, and training and supervision of professionals.

The chapter discusses several proposed models, some of which are available as research frameworks, prototypical solutions, or commercial systems, and others are still neglected or simply addressed as important for future research.

Many studies have been conducted, and many solutions have been proposed at different levels (from the clinical or from the AI points of view), but some critical aspects make it quite complex to analyze the literature; for example, from data privacy, to the evaluation of safety of therapeutic processes, to the multidisciplinary knowledge needed to model and evaluate clinical interactions, to the development of complex strategies, and to the evaluation of results, when not only micro-interventions are planned, etc.

Today, NLP and ML allow us to treat a huge amount of data, but clinical corpora are still not available or are quite limited, due to very restrictive therapeutic protocols and very specialized domains. Corpus samples are often too few to calculate robust evaluation figures and the developers often do not compare their results with others, in a satisfactory way (e.g., between frame-based, supervised, and unsupervised algorithms). Sometimes, authors use private corpora, or the technical specifications of the proposed systems are not accessible.

Human-like interaction – imitated or complemented by CAs – is a central aspect of the chapter: both from the psychotherapeutic and the CA, NLP, and ML points of view. In particular, it's clear the relevance of empathic-oriented behaviors, sentiments (and emotion analysis), prosodic and "mirroring" competencies, in the context of evidence-based protocols that try to ensure the efficacy of augmented psychotherapeutic strategies and the subjects' and therapists' adherence to them.

A multidisciplinary approach to those aspects will contribute to a new generation of (possibly embodied) CAs for psychotherapy, being able to orient the future research directions.

9.2 APPLICATIVE DOMAIN AND PSYCHOTHERAPEUTIC USE OF CAs

9.2.1 The PCC Approach, Ethics, and Safeness Using AI Solutions

Attention to the Patient-Centered Care (PCC) paradigm [Balint, 1969], to ethics principles, and to the safeness of interactions involving CAs (in particular, in a critical applicative domain like the mental health intervention), is orienting the most innovative research, system development, and their evaluation. The present chapter emphasizes those aspects, presenting different models and proposed solutions.

As a general overview, most proposed CAs address mood, anxiety, depression, Post-Traumatic Stress Disorder (PTSD), drug abuse, and Autistic Spectrum Disorder (ASD), while relatively few of them address schizophrenia, dementia, phobic disorders, psychosis, stress, eating disorders, obsessive-compulsive disorder, and bipolar disorder [Abd-Alrazaq et al., 2019, Abd-Alrazaq et al., 2020, Abd-Alrazaq et al., 2021]. Moreover, many referred clinical interventions are based on Cognitive Behavioral Therapy (CBT) and evidence-based methodologies.

AI solutions – involving frame-based, rule-based, ML statistical supervised or unsupervised algorithms, or, more recently, Deep Learning-based approaches – have been proposed with different clinical goals: diagnosis, prognosis, learning of skills, counseling, treatment, post-treatment reinforcement, detection and monitoring of potential biomarkers, results analysis, training of specialists, gender identity support, personal narrative encouragement, and perception and communication of intimate behavior. This chapter will present a reasoned selection of them, with a particular focus on NLP-based and ML-based autonomous CAs for psychotherapy and training.

9.2.2 Autonomous CAs for Psychotherapy

Despite the fact that, in the current public perception, the topic of new technologies in the field of mental health is fraught with reservation and fear [Bendig et al., 2019], research into the psychotherapeutic application of autonomous (embodied) CAs (sometimes called *chatbots*) is emerging:

> The field is characterized by a large variety in all its aspects, for example, type of intervention, target behavior, platform, ECA embodiment, communication modalities, 'ECA' mental states, and study design.

While there are several studies surpassing the development and piloting phases, as might be expected in a relatively new field, evidence about the clinical effectiveness of ECA applications remains sparse. Technologically advanced ECA applications are very interesting and show promising results, but their complex nature makes it difficult for now to prove that they are effective and safe to use in clinical practice.

At the present, their value to clinical practice lies mostly in the experimental determination of critical human support factors. In the context of using ECAs as an adjunct to existing interventions with the aim of supporting users, important questions remain with regard to the personalization of ECA's interaction with users, and the optimal timing and manner of providing support.

From Provoost et al. [2017]

Let us define two important concepts: *autonomy* and *embodiment* of the CAs when involved in psychotherapy.

- *Autonomous* CAs are applications that do not need any human intervention to interact with the user (in our case, the client of the therapist); they understand the user's sentences and are able to generate suitable responses according to some therapeutic criteria.
- The capacity of the CA to simulate human-like multimodal interaction is what makes it *embodied*.

But "human-like multimodal interaction" could mean quite different things. In fact, one could refer to the multimodality aspect (text, voice, facial expressions, body gestures, etc.) or to the human-like aspect (ability to manage emotions, empathy, mirroring, etc.) up to a (more or less) realistic virtual (or robotic) body, maybe enriched by esthetic attributes, personality, and interaction style.

Many of the previous points are important research topics. Understanding what is better to realize and imitate, and what is useless (or even perceived as threatening), with respect to the human-like aspect, is still discussed, in particular when dealing with psychotherapy.

From this point of view, it is important to note that *more is not always better*. For example [Vaidyam et al., 2019] agree with [Ardito and Rabellino, 2011] by saying that speech is more important for some clients to create empathic interactions than a 3D avatar. Moreover, note that the visual contact is treated quite differently by diverse psychotherapeutic models (e.g., the couch in psychoanalytical intervention avoids visual contact to facilitate the processing of unconscious content).

9.2.3 Guided/Unguided/Augmented Psychotherapeutic Interventions

As we will see, NLP and ML models can be used for realizing some components of CAs, but also for analyzing results and different intervention styles of psychotherapists, for supporting training and supervision of psychotherapists, and for providing real-time or post-session instruments to analyze human–human and human–machine interactions (to learn better methods for dialogue management, and more conversational flexibility and efficacy with different goals).

From Internet- and Mobile-based Interventions (IMIs) to CAs, while numerous studies found [Andersson et al., 2014; Carlbring et al, 2018 *quoted in* Bendig et al., 2019] that these interventions –often using cognitive-behavioral techniques – show comparable effectiveness to classical face-to-face psychotherapy, and that problems such as anxiety and depression can already be effectively addressed in this way, CAs capable of more complex interventions are still lacking (i.e., ones relying on other psychotherapeutic models, not only based on behaviors but also considering meanings or unconscious aspects of the narrative, or complex relational aspects as transfer and counter-transfer).

CAs are involved in psychotherapy in a *guided* or *unguided* way. The guided approach typically involves licensed health professionals and is usually more clinically effective than the unguided approach. Moreover, guided interventions have been discovered to improve adherence and thus the effectiveness of the psychotherapy [Provoost et al., 2017]. The authors underline that the unguided interventions (applications in which the CA is used as an adjunct to an intervention that could also have been used independently) are mostly CBT-based programs, educational aids, and self-management interventions, whereas guided interventions are mostly about training of social interaction skill, and counseling. More complex psychotherapeutic interventions, which globally and systematically preview the complementary roles of psychotherapist and CA, are now under study and evaluation, and are sometimes referred to as *augmented psychotherapy.*

9.2.4 Sentiment, Emotions, and Empathy: Richness of Therapeutic Interactions

From Picard's pioneering studies [Picard, 1997], much research has been done on *sentiment* and *affective* computing as a fundamental component of meaningful human interactions. Only recently more adequate models are trying to include *empathy,* which is the capacity to relate to another's emotional state, and which allows to enhance the interaction in different applicative domains such as education, gaming, training, companionship, and clinical interventions (the focus of this chapter).

As emphasized by Yalçin [2018] and showed by studies conducted so far, empathic CAs lead to more trust, are able to manage longer interactions, help cope with stress and frustration, and increase engagement.

New multilevel models of empathy [Asada, 2015; Morris et al., 2018; Yalçin, 2018] underline its role in processes involving the building and transformation of self in relation with others, personality, and personal and group well-being, in agreement with psychology, neuroscience, and models described by biologically inspired studies [Damasio, 2010, 2021; Panksepp and Biven, 2012; Panksepp and Davis, 2018].

But important questions remain and will motivate future research [Yalçin, 2018]: how can we model empathy in CAs, and what are the requirements/components for an empathic CA? How can an empathy model be simulated in an embodied CA? How can we evaluate an empathic CA? From the theoretical and the empirical background on empathy, Yalçin (among others) proposes to categorize the component of empathy as follows: emotional communication competence (emotion recognition, emotion expression, and emotion representation), emotional regulation (self-related modulation factors as mood and personality, and relationship-related modulation factors as affective link, liking, and similarity), and cognitive mechanisms (appraisal and re-appraisal, theory of mind, simulation theory, and self and other oriented perspective-taking). With the goal of creating a dynamic empathic CA able to interact with the user in real time, Yalçin also underlines the importance of considering the emotional communication competence as the foundation of empathic behavior and how it changes the perception of empathy during interaction. AI interactive systems use the answers to these questions to develop CAs able to act empathically and to evoke empathic responses in the user.

Empathic CAs in psychotherapy are actually studied by many researchers, both from the clinical and the technological points of view, as interesting and challenging frontiers of AI, NLP, and ML. Even if truly empathic CAs are not yet available, many studies contribute to model and implement ML-based CA sub-components devoted to emotion, intention recognition, complex goals achievement, and relational syntonic modulation, allowing a richer evaluation of the involved innovative solutions. Between them, the already quoted [Yalçin, 2018] and [Morris et al., 2018] and the last studies published by the author's NLP laboratory members [Scotti et al., 2021; Scotti, 2023].

9.3 CLINICAL, TECHNICAL, AND USER EXPERIENCE EVALUATION

The complexity of the considered domains reflects on evaluation complexity, where clinical, technical, and user experience evaluation need to be measured in a quantitative and/or qualitative way. Generic evaluation criteria of CAs – such as *practicability, feasibility,* and *acceptance* – should be extended with *effectiveness, sustainability,* and especially *safeness* [Bendig et al., 2019].

As underlined by Provoost et al. [2017], on one hand, usual principles used to evaluate CAs – such as the *engagement* with the agent – shouldn't be considered enough; on the other hand, clinical specific principles – such as treatment adherence (of the patient and/or therapist) – could be too simplistic when not considered in connection with effectiveness. Furthermore, many CAs for therapy are still in a piloting phase and more complex evaluation methodologies have to be defined to respect the requirements of the UK Medical Research Council (MRC) framework for complex interventions [Craig et al., 2008].

Laranjo et al. [2018] underline the heterogeneity in evaluation methods and measures and the predominance of quasi-experimental study designs over Randomized Controlled Trials (RCTs). Most of the research in the area evaluate task-oriented CAs that are used to support patients and clinicians in highly specific processes. The only RCT evaluating the efficacy of a CA found a significant effect in reducing depression symptoms [Fitzpatrick et al., 2017]. Two studies comparing diagnostic performance of CAs and clinicians found acceptable sensitivity and specificity [Philip et al., 2014; Philip et al., 2017].

The evaluation of efficacy in reducing symptoms, sensitivity, and specificity are often obtained with the heterogeneity of evaluation methods and measures, and the predominance of quasi-experimental study designs over RCTs. Most research studies evaluate task-oriented CAs that are used to support patients and clinicians in highly specific processes. Three types of evaluation have been adopted and guided the selection of the analyzed papers in Laranjo et al. [2018]:

- *Technical evaluation*: objective assessment of the technical properties of the CA as a whole, and evaluation of its individual components (e.g., reported figures of technical performance such as the proportion of successful task completions and the recognition of accuracy.)

- *User experience:* the overall user satisfaction is usually evaluated, but the properties of components can be asserted too by means of qualitative (e.g., focus group) or quantitative (e.g., survey) methods.

- *Health research evaluation*: health-related results are presented in the study, including process and outcome measures. For example, effectiveness in symptom reduction, diagnostic accuracy, narrative skills, mental health symptoms disclosure, behavior change, and adherence to self-management practice (via qualitative methods).

In addition, to fully describe a CA, other important characteristics should be reported by Provoost et al. [2017]:

- *Development phase*: development, piloting, evaluation, and implementation.

- *CA's communication modalities*: speech, facial and gaze expressions, hand and body gestures, text, and touch.

- *User's communication modalities* (that CA is able to detect): speech, facial and gaze expressions, hand and body gestures, text, and touch.

- *Personalization*: static user model, dynamic user model, menu-based dialogue, and natural language dialogue.

- *Platform*: serious game, stand-alone software, robotic, virtual reality, and web-based.

- *Interventions*: social skill training, educational aid, psychotherapy, CBT, counseling, and self-management.

- *Social role*: social interaction partner, tutor, coach, and health care provider.

- *Outcomes*: non-clinical (usability, satisfaction, usage) and clinical (behavioral, self-report, knowledge).

- *Study participants*

To improve reporting of studies, and to enable readers to assess the quality of studies, combine results and interventions, Laranjo et al. [2018], and Abd-Alrazaq et al. [2019] recommend following standards such as CONSORT-EHEALTH [Eysenbach, 2011], TREND [Des Jarlais et al., 2004], and STARD [Bossuyt et al., 2015]. Furthermore, Vaidyam et al. [2019] underline that the WHO mHealth Evidence Reporting and Assessment (mERA) framework [Agarwal et al., 2016] should help to choose future CA attributes that currently lack consensus. This could contribute to avoid another problem related to validation: many health-related CAs on the market have not been empirically validated [Bendig et al., 2019]; the certification of therapy-relevant CA as "medical device" could resolve this problem, as medical devices need to obtain the CE (*Conformitée Européenne*) label [Rubeis and Steger, 2019 *quoted by* Bendig et al., 2019].

Finally, the Aafjes-van Doorn et al. [2020] review – focused on ML-based CAs and frameworks for psychotherapy – underlines further important aspects that should be considered for evaluation and reporting. Quoting from it and analyzed studies, the following:

1. The sample size: ML algorithms require larger sample sizes than traditional statistical methods but exactly how large remains unclear (and discussed). If a relatively small number of participants is involved in experiments, there is the risk that the model will be overfitted, that is, specific of that dataset, and not accurate in making predictions in a new dataset.

2. The model performance: is it possible to compare performances of ML based modules with traditional statistics ones? Should they be both applied to identify different variables? Are numeral results from both directly comparable? The discussion is open [Atkins et al., 2012].

3. The evaluation metric and what is considered an acceptable level of accuracy depends on what the algorithm will be used for (i.e., in predicting suicidal ideation, one might want to have a model that minimizes false negatives over minimizing false positives).

4. As is true for any ML application, the size and quality of the data limit model performance [Graham et al., 2019]. ML internal and external cross-validation should be done and specified.

5. Assessing big data confidentiality and protecting identities need research: with the large samples sizes and range of data types and sources, identities can be reconstructed by combining pieces of information, each of which would not be enough to identify a person but, combined, would allow individuals to be identified [Berman, 2013].

6. Currently, there appears to be a lack of guidance on development of ML applications, their clinical integration and training of psychotherapists, as well as a 'gap' in ethical and regulatory frameworks [Fiske et al., 2019]. Institutional Review Boards may also have limited knowledge of emerging ML methods and applications, which makes risk assessment inconsistent.

7. Interpreting multiple latent variables (e.g., in deep learning) is complicated, and more work is required to bridge the gap between ML in psychotherapy research and clinical care. Accessible ML education and tool development is required to facilitate understanding and usage in the wider clinical research community.

From Aafjes-van Doorn et al. [2020]

9.4 SURVEYS, SCOPING REVIEWS, AND FIRST SYSTEMS FOR PSYCHOTHERAPY

9.4.1 Recent Scoping Reviews

Papers referring to the emerging usage of CA for mental disorders have been published in different domains (from psychology to computer science, from medicine to interdisciplinary databases, etc.), making it quite difficult to collect a specific literature. Thus, recent scoping reviews are particularly useful; Provoost et al. [2017]; Laranjo et al. [2018]; Bendig et al. [2019]; and Vaidyam et al. [2019] will be presented in this section, while Aafjes-van Doorn et al. [2020] in the following one. Such reviews map – in different moments and from different points of view – the key concepts underpinning the AI and CA research area, when applied to mental health and clinics, and the main sources and types of evidence that are available. They address broader topics where many different study designs might be applicable, and do not emphasize quality assessment of the included studies.

In Provoost et al. [2017], 54 studies have been analyzed. More than half of them (26) focused on autism treatment, and CAs were mostly used for social skill training. This emphasized the usage of CAs as social interaction partners reinforcing or introducing lacking social behaviors, such as joint attention, communication, tactile interaction, imitation, turn taking, and job interview skills. More generally, applications ranged from simple reinforcement of social behaviors through emotional expressions to sophisticated multimodal conversational systems.

Most applications were still in the development and piloting phase and not yet ready for routine practice evaluation or application. Few studies conducted controlled research on clinical effects of CAs, such as reduction in symptom severity. Even if, in most cases, clinical behavioral outcomes were restricted to just pre-/post-measurements within experiments that involved relatively small sample sizes, CAs showed a positive effect on user engagement and involvement, in particular when CAs were adopted as adjunct to already existing CBT-based interventions for different mental disorders (depression, anxiety, PTSD).

Despite the issues discussed above, the analyzed papers permit to draw some interesting conclusions: there is a steady growing number of papers per year, from 2010, dedicated to "CA and clinical interventions"; CAs often show a rich set of multimodal communication modalities, while the modalities allowed to users are usually quite simple; and many papers emphasize the importance of clinical and non-clinical outcomes. Moreover, in particular treating depression and anxiety disorders, the anonymity of CAs, their availability, non-judgmental nature, and the ability for people to practice social interaction in a safe environment, were confirmed to be important reasons to use CAs.

The central role of NLP and AI algorithms appears evident in the second scoping review published in 2018. The systematic review protocol of Laranjo et al. [2018] allowed to include 17 studies evaluating 14 different CAs with unconstrained natural language input capabilities. CAs were supported by different technologies (apps delivered via mobile device, web or computer, SMS, telephone, and multimodal platforms). They were task/non-task oriented, with a mixed/system/user dialogue initiative (some working in the domain of depression, anxiety, and PTSD), with different input and output formats.

The following are names of a few CAs: Woebot [Fitzpatrick et al., 2017], Miner et al.'s study [Miner et al., 2016b], Harlie the Chatbot [Ireland et al., 2016], Bzz [Crutzen et al., 2011], and SimSensei Virtual Agent [Lucas et al., 2017].

Most were implemented by means of finite-state, frame-based approaches, while just one was able to manage complex dialogues [Miner et al, 2016b]. The Bendig et al. [2019] scoping review only considered studies that describe CAs based on evidence-based clinical psychological and psychotherapeutic scripts. The review included six papers, all published between 2017 and 2018; five of them were based on cognitive-behavioral scripts and focused on depression, anxiety, mental well-being, and stress. To exemplify the complexity of dialogues, the authors describe the design of their chatbot named SISU [Bendig et al. 2019].

Here are the CAs described in the review: Mylo [Bird et al., 2018], Woebot [Fitzpatrick et al., 2017] Gabby [Gardiner et al., 2017], Shim (Ly et al., 2017), PEACH [Stieger et al., 2018], and Sabori [Suganuma et al., 2018]. The six pilot studies mainly concerned with evaluating the practicability, feasibility, and acceptance of these chatbots. Unfortunately, the datasets employed to evaluate the effectiveness (with respect to well-being, stress, and depression) are often too small for providing high-quality statistical figures and thus a reliable assessment.

Finally, Vaidyam et al. [2019] focused on populations with or at high risk of developing depression, anxiety, schizophrenia, bipolar, and substance abuse disorders, and considered studies that involved CAs in a mental health setting. The ten studies that met inclusion criteria confirmed the potential of CAs in psychoeducation, self-adherence, high satisfaction rating. In some cases, performances on diagnoses were declared as good, and the risk of harm from the use of CAs as very low. Self-care pathways were described in both clinical and non-clinical populations.

Vaidyam et al. [2019] focused on some of the studies considered by the previous surveys. First of all, as previously mentioned, they underlined that the voice is the most important factor of a positive experience with a CA [Ardito and Rabellino, 2011]; moreover, although the therapeutic relationship establishment early in traditional therapy, is predictive of favorable outcomes, little is today known regarding how patients feel supported by CAs and how this relationship develops and affects psychiatric outcomes. On the latter point, Fitzpatrick et al. [2017], Gardiner et al. [2017], and Bickmore et al. [2010] highlight the effect of establishing appropriate rapport or therapeutic alliance on patient interactions. Scholten et al. [2017] and Bickmore and Gruber [2010] suggest patients may also develop transference towards CAs, leading to unconscious redirection of feeling towards CAs.

As final comments on the surveys presented above, Provoost et al. [2017] stress that an adaptable, trustworthy, "guiding rather than directive" coaching role, and the capacity of empathic expressions without reflecting negative ones back to the users, are all desired quality of a CA that try to address empathy and emotional states during clinical interactions. Vaidyam et al. [2019] underline that creating CAs with empathic behavior is an important research area. Scholten et al. [2017] further state that alliance is better formed between patients and chatbots with relational and empathetic behavior, suggesting that patients may be willing to interact with these CAs even if their functionalities are limited and even if they know that a CA cannot really empathize with "lived experiences." Moreover, another important factor is the unconstrained availability of CAs, which creates the opportunity for therapeutic sessions whenever the patient wants and needs them [Vaidyam et al., 2019]; however, the effectiveness of the therapy could also be negatively affected by this "always available" support, which poses the risk for the patient to become overattached or even codependent.

9.4.2 Woebot and Shim: Two CAs for Mental Health and Wellbeing

Woebot [Fitzpatrick et al., 2017] is a platform-independent app designed for therapeutic use (its CA can be deployed on message systems such as Facebook Messenger, Kik, Twitter) which has been designed to present mental health materials in an interactive and conversational style. It is frame-based, supports a mixed dialogue initiative, and is limited to textual input/output. It is a non-embodied CA, and the name has been chosen to emphasize the non-human nature of the agent. Finally, it is unable to reflect any deep understanding of the user's particular situation.

The paper by Fitzpatrick et al. [2017] is one of the most cited contributions in the field of CAs for psychotherapy. In particular, the objective of the study was to determine the feasibility, acceptability, and preliminary efficacy of a fully automated CA to deliver a self-help program (based on CBT) for college students who self-identify as having symptoms of anxiety and depression. Ethics and Informed Consents have been considered: the study was reviewed and approved by Stanford School of Medicine's Institutional Review Board.

Many publications on Woebot (some of them peer-reviewed, others under approval) can be retrieved on the web, describing experiments of controlled usage and evaluation of the agent, and attesting the evidence of clinical efficacy, declaring Woebot to be a feasible, engaging, and effective way to deliver CBT.

Fitzpatrick et al. [2017] describe their CA as follows:

The bot's conversational style was modeled on human clinical decision making and the dynamics of social discourse.

Psychoeducational content was adapted from self-help for CBT [...]. Aside from CBT content, the CA was created to include the following therapeutic process-oriented features:

- Empathic responses: The CA replied in an empathic way, appropriate to the participants' inputted mood...

- Tailoring: Specific content is sent to individuals, depending on mood state...

- Goal setting: The CA asked participants if they had a personal goal that they hoped to achieve over the 2-week period.

- Accountability: To facilitate a sense of accountability, the CA set expectations of regular check-ins and followed up on earlier activities, for example, on the status of the stated goal.

- Motivation and engagement: To engage the individual in daily monitoring, the CA sent one personalized message every day or every other day to initiate a conversation (i.e., prompting). In addition, "emojis" and animated gifs with messages that provide positive reinforcement were used to encourage effort and completion of tasks.

- Reflection: The CA also provided weekly charts depicting each participant's mood over time. Each graph was sent with a brief description of the data to facilitate reflection...

From Fitzpatrick et al. [2017]

The authors realized a Randomized Control Trial. In an unblinded trial, 70 individuals (age 18–28 years) were recruited online from a university community social media site and were randomized to receive either 2 weeks (up to 20 sessions) of self-help content using Woebot, or were directed to the National Institute of Mental Health ebook, *Depression in College Students*, as an information-only control group.

All participants completed web-based versions of the 9-item Patient Health Questionnaire (PHQ-9), the 7-item Generalized Anxiety Disorder scale (GAD-7), and the Positive and Negative Affect Scale at baseline and 2–3 weeks later (T2).

Acceptability and usability were tested with qualitative questionnaires, too:

> Two major themes emerged as 'the best thing about the experience using Woebot': process and content. In the process theme, the subthemes that emerged were accountability from daily check-ins (noted by 9 participants); the empathy that the bot showed, or other factors relating to his "personality" (n=7); and the learning that the bot facilitated (n=12), which in turn was divided into further subthemes of emotional insight (n=5), general insight (n=5), and insights about cognition (n=2).
>
> Three themes emerged as 'the worst thing about the experience using Woebot': process violations (n=15), technical problems (n=8), and problems with content (n=8). By far the most common subtheme to emerge among the process violations related to the limitations in natural conversation such as the bot not being able to understand some responses or getting confused when unexpected answers were provided by participants (n=10), and 2 individuals noted that the conversations could get repetitive. Technical problems were described by 8 individuals, with technical glitches in general (n=4) and looping conversational segments (n=4) emerging as subthemes. Problems with content were described by 8 individuals, with most of these relating to emoticons and either interactions or content length 8.
>
> *From Fitzpatrick et al. [2017]*

Presenting the results of this preliminary trial, the authors suggest important aspects to be considered in the future. Considering that Bickmore et al. [2005] demonstrated that individuals can develop a measurable therapeutic bond with the CA after 30 days of usage, the authors underline that a standardized measure of working alliance should be explicitly explored using – for example – the Working Alliance Inventory [Horvath and Greenberg, 1989]. Furthermore, the study suggests that CA process factors, such as the ability to convey empathy, may be capable of both amplifying and, conversely, violating a therapeutic process. This underscores the importance of including trained and experienced clinicians in design of clinical app processes.

Shim [Ly et al., 2017] is a fully automated CA built as a smartphone app, which has been designed to deliver strategies used in positive psychology and CBT interventions for a non-clinical population. It interacts and conducts conversations via vocal or textual methods.

The study has been designed to assess the effectiveness and adherence of Shim as well as to explore participants' views and experiences of interactions with the CA. It was designed as a randomized controlled study in a non-clinical population, comparing outcomes from two weeks' usage of the positive psychology-oriented Shim, against a waitlist control group.

The authors underline the design strategies: the goal of the conversations in Shim is to help the user reflect upon, learn, and practice these small strategies and behaviors.

> The conversations in Shim are centered around insights, strategies and activities related to the field of positive psychology. These include (but are not limited) to expressing gratitude, practicing kindness, engaging in enjoyable activities and replaying positive experiences. Also, components from the third wave of CBT are included in the strategies taught by Shim, such as present moment awareness, valued directions and committed actions.

> The dialogues in Shim have been pre-written by professionals with training in psychology. Each dialogue can be represented as a tree graph with one or multiple starting points and one or multiple closing points. The user replies to Shim mostly via two types of statements: (1) comments, requiring inputs from the user such as free text, (2) picking elements either from a list or from a fixed set of reply options. With language pattern matching and keyword spotting, as well as conditional expressions, Shim gives adequate responses to the user's statements.

> *From Ly et al. [2017]*

The outcome figures provided by the authors were: the Flourishing Scale (FS), a brief 8-item summary measure of the respondent's self-perceived success in important areas such as relationships, self-esteem, purpose, and optimism [Diener et al., 2009]; the SWLS, is a 5-item self-report,

7-point scale concerning subjective well-being [Diener et al., 1985], which is assessed by measuring cognitive self-judgment about satisfaction with one's life; and the PSS-10, which measures the perception of stress (i.e., degree to which situations are appraised as stressful), by asking respondents to rate the frequency of their thoughts and feelings related to situations that occurred in recent time [Cohen et al., 1997].

The fully automated intervention allowed significant effects on psychological well-being (FS) and perceived stress (PSS-10) among participants who adhered to the intervention, when compared to the waitlist control group. The participants showed high engagement during the 2-week-long intervention. Both these aspects are higher if compared with other studies on fully automated interventions claiming to have a good level of user engagement.

Furthermore, qualitative questionnaires revealed interesting involved sub-themes, with respect to *Content* (Activation, Learning, Reflection, Repetitiveness, Shallowness), *Medium* (Routine, Availability, Moderator, Lack of Clarity), and *Functionalities* (Weekly Summary, Lack of Notification, Restricted User Interface).

9.5 ML-BASED SOLUTIONS AND EMPATHIC CAs FOR PSYCHOTHERAPY

9.5.1 Recent Survey on ML-Based CAs and Frameworks for Psychotherapy

Aafjes-van Doorn et al. [2020] presented the literature on ML applications in mental health and psychology, focusing on psychotherapy research. During the previous two years, other scoping reviews concentrated on ML approaches in psychiatry, mental health, and psychology, but no specific selection was done on psychotherapy. In particular, Dwyer et al. [2018], Shatte et al. [2019], and Graham et al. [2019] focused on ML when used for diagnosis, prognosis, treatment, prediction, detection and monitoring of potential biomarkers; Imel et al. [2017] – which will be deeper analyzed in section 9.6.2 – underlined that technology-enhanced human interaction, including ML, is likely to have a significant impact on mechanism and process, training and feedback, and technology-mediated treatment modalities; finally, Tai et al. [2019] suggested that ML can be used in unison with psychiatry by analyzing the multidimensional, multilevel disease models helping mental health practitioners redefine mental illness more objectively than currently done in DSM-5.

Aafjes-van Doorn et al. [2020] selected 51 studies and analyzed them to inform clinicians in the methods and applications of ML in the context of psychotherapy, to clarify the strengths and weaknesses of these methods and considerations within psychotherapy research, and to highlight clinical implications and identify potential opportunities for further research.

Two types of studies could be distinguished from the 51 studies: 47 focused on developing and testing their own ML model, while seven [Burns et al., 2011; Hirsch et al., 2018; Imel et al., 2019; Inkster et al., 2018; Krause et al., 2019; Tanana et al., 2019; Watts et al., 2014] reported on the feasibility of treatment tools that use already-existing ML algorithms. Most of such 47 studies used supervised learning techniques to classify or predict labeled treatment process or output data, whereas others used unsupervised techniques to identify clusters in the unlabeled patient or treatment data. Moreover, 16 applied NLP analysis to transform raw texts into more useful labeled data.

Among the remaining seven studies, Tanana et al. [2019] and Hirsch et al. [2018] will be deeply analyzed in Section 9.6.2.

The results of the survey are reassumed in the following:

> There was considerable heterogeneity in the nature of how the results were reported across studies.

> 12 studies examined whether ML models could be used to predict behavioral or observational codes (ratings/labels) assigned by human experts; 5 studies examined ML models to identify characteristics of sessions (transcripts of texts) that predict outcomes (end-of-treatment or within sessions); 13 studies used ML models to predict treatment outcome based on pre-treatment or questionnaire/intake data; 9 studies predicted treatment outcome based on neuro-imaging data; 4 studies demonstrated the use of ML analytics for linguistic coding (between them Imel et al. [2019]); 2 studies used ML models to predict treatment outcome based on ecological momentary assessment during treatment. (...)

> Most of the 44 reviewed studies concluded that ML models were effective in predicting the target, whether it was human codes used to label data or treatment outcomes, and implied that the ML approach was more beneficial than previously applied traditional statistical approaches. However, as described above, the level of

accuracy, sensitivity, or specificity that is considered to be acceptable varies depending on the aims of the study and the dataset. None of the studies explicitly compared the ML performance with that of more traditional statistical analyses. (…)

Caution is necessary in order to avoid over-interpreting preliminary results. (…)

Overall, the current applications of ML in psychotherapy research demonstrated a range of possible benefits for identifications of treatment process, adherence, therapist skills and treatment response prediction, as well as ways to accelerate research through automated behavioral or linguistic process coding. Given the novelty and potential of this research field, these proof-of-concept studies are encouraging, however, do not necessarily translate to improved clinical practice (yet).

From Aafjes-van Doorn et al. [2020]

The analyzed studies show that ML brings new possibilities for analyzing larger datasets, but also that further clinical-research collaborations are required to fine-tune ML algorithms for different treatments and patient groups, and to identify additional opportunities for ML applications to advance psychotherapy process and outcome. In particular, ML could help therapists identify mental illnesses at an earlier stage, understand when and how interventions can be more effective, personalize treatments based on an individual's unique characteristics, and focus on the relational aspects of psychotherapy that can only be achieved through therapist–patient interactions. To emphasize the role of ML in psychotherapy, some of the last points will be further described in the following section.

9.5.2 Three ML-Based Systems: Adikari et al. Framework, Wysa, and PopBots

NLP and ML techniques that are able to improve emotional understanding and empathetic behaviors of CAs in psychotherapeutic relationships characterize the most recent systems analyzed in this section.

The rich and original contribution of Adikari et al. [2019, 2022] is based on a complex cognitive model of users and interactions. The empathic CA framework for real-time monitoring and co-facilitation of PCC is well described in the paper. The approach uses NLP and AI algorithms to detect patient emotions, predict patient emotional transition, detect group emotions (based on individual emotions and collaboration metrics), and formulate patient behavioral metrics (based on active, passive participation information, and emotional support metrics).

An ensemble of NLP and ML techniques (among them, Finite State Machines and Markov models, lexicon-based models, Naïve Bayes, Random Forest, Support Vector Machines, Multilayer Perceptron, and Logistic Regression) have been used to train, classify, predict, and extract shared contents.

To derive a score from patients' behavior, the authors combined two metrics (*emotion engagement* and *participation*):

- First, in order to derive the 'emotion engagement score' of each patient, we use emotions, group emotion mentions captured by emotion extraction components. Besides, we capture specific concerns expressed by patients based on the domain and theme of discussion. As a baseline version, we used a clinical ontology of patient concerns provided by the clinicians and therapists. This included several physical symptoms and social concerns often faced by patients. The concern list was also further enriched by using the trained word2vec models, which identified similar expressions used by people to express different concerns. Following these extraction steps, the relative number of high emotional posts, group emotion and concern mention posts by each patient were used to derive the 'emotion engagement score' metric.

- Second, in order to capture the 'participation score' we calculated the volume of content shared by the patient using the average length of posts and the average number of posts within a conversational setting. Based on these data, we introduce a quantified measure to assess patient behaviors that encompass the information and participation score of each patient to represent the volume of content shared by the individual. The two measures were combined using the fuzzy integral. The scores for each metric were combined using a fuzzy measure to derive the final patient behavioral score.

From Adikari et al. [2022]

It is important to underline the capacity of the framework to make proactive decisions based on the patient's emotional state, and generate personalized responses based on the emotional characteristics of each individual. The CA's automated response generation is triggered based on multiple factors of the patient behavior, such as a probability to transition into negative emotions and prediction of imminent negative emotions. A rule-based emotional message generation, based on the "negativity threshold" of each patient is used by the systems. Individuals who show a higher-than-usual propensity to transit to negative emotion states, based on the transition matrix and predictions of imminent negative emotion, are notified to the healthcare practitioners as well as used for automated response generation.

Based on patient behavior, the CA supports the psychotherapist with specific algorithms for the propagation of empathic conversational elements to facilitate new strategies for the human therapist or care provider.

The validation of effectiveness, practical value, and core capabilities of the framework, was based on a clinical protocol relying on an online, professionally led, synchronous, text-based online support group for cancer patients and caregivers across six provinces in Canada.

The authors, finally, underline that emotion extraction could be improved in the future by addressing the current limitations related to detecting ambiguous expressions, idioms, and indirect emotion statements (i.e., associated to figurative language and irony).

The second system we will consider is Wysa, a Smartphone-Based Empathetic Artificial Intelligence Conversational Agent promoting mental resilience and well-being using a text-based conversational interface. Wysa is declared as a non-clinical device, but it has been proposed to young persons with self-reported symptoms of depression (different contexts were considered, among them many university campuses) involving and teaching them how to manage their anxiety, energy, focus, sleep, relaxation, loss, worries, conflicts, and other situations.

Despite difficulties encountered in retrieving technical information on the architecture and models adopted by Wysa, interesting analysis of its clinical usage could be found in Inkster et al. [2018], Meadows et al. [2020], and Beatty et al. [2022].

In Inkster et al. [2018], two groups of users (*high users* and *low users*) who engaged in text-based messaging, and self-reported symptoms of depression using the PHQ-9, were observed using a quantitative and a qualitative approach with good results (but in a too-short time, and involving a too-small sample set).

The described quantitative analysis measured the CA's impact by comparing the average improvement in symptoms of depression between high and low users. An impact (pre-post) analysis was conducted in relation with a context/descriptive one.

The qualitative analysis measured the CA engagement experience (from both effectiveness and efficiency points of view) by analyzing in-app user feedback and evaluated the performance of a ML classifier to detect user objections during conversations. An NLP-based qualitative thematic analysis, on in-app feedback responses was performed.

The Wysa scientific advisory boards helped to create the capability of the app to respond to emotions that a user expresses over written conversations. They helped to design contents and tools based on evidence-based self-help practices such as CBT, dialectical behavior therapy, motivational interviewing, positive behavior support, behavioral reinforcement, mindfulness, and guided microactions and tools to encourage users to build emotional resilience skills.

Meadows et al. [2020] compare Woebot, Wysa, and Tess [Fulmer et al., 2018; Joerin et al., 2019] in sessions lasting a few weeks, to underline that the traditional concept of mental health "recovery," in technologically augmented pathways is going to become a cooperative "walkthroughs" process. The role of AI technologies and the temporal articulation to adopt were also considered. These aspects affect the evaluation parameters, the app's vision, its operating model, governance, mediator characteristics, registration and entry, and suspension and closure.

Beatty et al. [2022] specifically concentrated on therapeutic alliance; however, their conclusions and use of this concept could be challenged by some specialists in the field, given that the evaluation was realized involving users for sessions lasting just a few days.

NLP-based processing is used by Salman et al. [2021] to analyze the empathic behaviors of two CAs used with kids: Dr Evie (eVirtual Agent for Incontinence and Enuresis) and SAM (Sleep Adherence Mentor). From psychological theories, 16 items were identified to analyze the empathic dialogic behavior of the CAs playing different roles (senior doctor, psychotherapist, nurse, or physician).

In particular, ten of the 16 items analyzed verbal relational cues to analyze empathy [Bickmore et al., 2005]:

- Empathy as a cue
- Social dialogue

- Reciprocal self-disclosure

- Metarelational communication

- Expressing happiness to see the user

- Talking about the past and the future together

- Continuity behaviors

- Reference to mutual knowledge

- Inclusive pronouns and politeness

- Greeting and farewell rituals

The remaining six verbal relational cues come from [Richards and Caldwell, 2017]:

- Motivational/encouraging adherence/confirming language/affirming language

- Decision making/empowerment/clarifying consequences/giving options

- Everyday conversational dialogues

- Information dialogues/educative/explanation

- Tasks/previous treatments/current health status/medical history/treatment adherence/recommendations (future treatments)/family history

- Empathy as reciprocal physical, emotional, and cognitive status

A qualitative analysis of the interactions was used to detect the psychology-based 16 relational cues. Statistical analysis was used to compare usage of relational cues among different healthcare roles.

Dr Evie is based on scripted dialogues, whereas SAM relies on a more sophisticated AI technology that encourages adherence, allows clarification cues, and considers the user's goals and beliefs to increase empathy. The architecture of SAM and the availability of interactions also allow preferences, medical history, contextual features, and personalization to be included in the CA's reasoning [Abdulrahman and Richards, 2019].

An interesting and different approach is described in Mauriello et al. [2021]. The authors developed PopBots, a fully automated mobile suite of shallow CAs (simplified CAs) for Daily Stress Management, not created to replicate or replace humans (i.e., family, friends, or therapists) but rather to operate as part of a larger ecosystem of agents providing stress management support.

In 2014, Paredes et al. [2014] demonstrated that a set of shallow CAs – when coupled with a web-based learning recommendation system – could help users to improve their long-term stress coping skills. Mauriello et al. [2021] extended the previous research on micro-interventions, exploring a suite of diverse and specialized shallow CAs for daily stress management, to demonstrate how new strategies may offer benefits for both users and designers:

> (1) multiple shallow chatbots are capable of delivering micro-interventions, lower barriers of time and commitment for users; (2) they can be authored and curated more quickly by novice designers to produce a variety of high-quality advice options; (3) this variety of chatbots could help improve long-term engagement (i.e., chatbots that fail could be removed); and (4) the suite approach allows for future personalization.
>
> *From Mauriello et al. [2021]*

To support the initial idea, they randomly assigned to each participant a different CA designed on the basis of a proven cognitive or behavioral intervention method, and then measured the effectiveness of such CAs, using self-reported psychometric evaluations of the participants' stress level (e.g., web-based daily surveys and PHQ-4).

Shallow CAs could be quickly developed and evaluated through a mixed methods exploratory study and then re-developed. Many conversations related to micro-interventions were thus obtained.

As a result, the authors' contributions include design recommendations:

- *Focus on lowering barriers to authorship and generating numerous shallow CAs based on the vast amount of available psychological interventions for stress management.*

- *Design for learning algorithms to handle recommendation and curation of interventions.*

- *Attempt to score, rank, and classify daily stressors before assigning CAs (interventions) to accommodate the differences in low- and high-complexity stressors as well as concerns about identifying problems that are too severe for the system to handle.*

- *Consider a multitude of user coping and conversational styles, including users who may need a guided intervention or just an opportunity to reflect by talking or typing it out into the void.*

- *Measure user personality, CA efficacy, and system engagement to optimize interactions across users.*

From Mauriello et al. [2021]

The authors hope that, enriched with a web-based learning recommendation system, PopBots could be used by the public health system.

9.6 MODELS AND FRAMEWORKS TO SUPPORT PSYCHOTHERAPIST TRAINING AND SUPERVISION

9.6.1 Conversational AI for the Analysis of Therapeutic Interactions and Relations

The Miner et al. [2016a], Miner et al. [2016b], and Miner et al. [2019] studies show that Conversational AI should be incorporated in psychotherapy for analyzing therapeutic interactions to foster innovative technological solutions and clinical interventions, extending training and supervision opportunities for psychotherapists.

Miner et al. [2016b], in particular, focuses on the Relational Frame Theory (RFT), an evidence-based theory of language and mental health that underlines the role of affects in language and considers the relation between sensation, affect, language, and behaviors.

To assess the effectiveness of CAs, the authors compared the sentiment dynamics in both human–human and human–CA dialogues. They found a persistence in human–human sentiment-related interaction norms when switching to human–CA dialogues, showing a tendency of users to respond positively to CAs. Some differences were found, however; for example, humans were twice as likely to respond negatively when faced with a negative utterance by a CA than in comparable situation with humans. Similarly, inhibition towards use of obscenity was greatly reduced. The authors emphasized that "*what makes a therapeutically successful conversation may be dramatically different from a non-therapeutic conversation...".* They analyzed the Fisher English Training transcript collection of 11,600 telephone conversations between human participants (corpus Fisher11k), and they used VADER [Hutto and Gilbert, 2014] to classify each conversation turn in Fisher11k into positive, negative, and neutral (VADER uses five grammatical and syntactical rules, and a lexicon that extends the Linguistic Inquiry and Word Count (LIWC) lexicon to cover micro-blogs) observing that:

> The Fisher conversations show a very strong tendency for participants to formulate positive-sentiment statements. Upon encountering negative statements, the participants showed a consistent tendency towards moving the conversations in a positive direction. Interestingly, this observation may identify a pattern that would not be clinically useful and could differentiate between non-therapeutic and therapeutic interactions.
>
> *From Miner et al. [2016b]*

To better analyze sentiments, the authors used two methods: a keyword-based approach able to identify sentiments (positive vs. negative moods that can be associated to events positive or negative elicitation), and a predictive model for the affects given a pair of dialogue lines (usually, a CA line followed by the user's replay). Such model was based in a RNN called Affective Neural Network with 2 Gated Recurrent Unit layers [Chung et al., 2014] and a Softmax activation function on the output layer. The model was trained leveraging emoticons that human partners embedded in their utterances to the system, mapping them to seven different affects: *anger, surprise, happiness, love, sadness, disgust, laughter.* They chose these affects starting from three sources: manually observed emotions generally exhibited by Cleverbot users, the set recently suggested by Facebook (*love, haha, wow, sad, angry*), and the six basic emotions suggested by Dr. Paul Ekman (*happiness, surprise, disgust, anger, fear, sadness.*).

Considering that awareness of affect mirroring is a key construct in successful therapeutic interaction, the authors connected RFT's concepts to NLP methodologies with a particular attention to empathy (connection and engagement). They, in particular, focused on how humans *mirror sentiment* and how a *reflective and validating language* is used by the therapist. The same can be said for reflective and validating communication that displays empathy.

The Miner et al. [2016b] study focuses on some well-known issues of NLP (e.g., analysis of sarcasm, with its mixture of multiple emotions) and remembers that language is just one "channel" humans leverage to express and understand emotions and empathy. Thus, integrating textual-based approaches with time and contingency aspects, or with other areas of affective computing (voice, gaze, facial expression, etc.), should benefit (as we will underline in Section 9.7) the design of mental health-focused CAs.

As always, with the goal to understand successful conversation strategies and to make use of these insights in counselor training, Althoff et al. [2016] concentrated on counselor conversations. Considering that previous studies in psycholinguistics demonstrated the words people use in therapeutic discourses can reveal important aspects of their social and psychological worlds [Beck, 1967; Pennebaker et al., 2003; Pestian et al., 2012; Ramirez-Esparza et al., 2008], Althoff et al. applied large-scale studies of Computational Linguistics to conversations in various clinical settings, including psychotherapy. They revealed subtle dynamics in conversations, such as coordination or style matching effects, social power and status, success in the context of requests, user retention, and novel styles.

Furthermore, they considered what Howes et al. [2014] wrote on psychotherapeutic interventions: symptom severity can be predicted from transcript data with comparable accuracy to face-to-face data, but insights into style and dialogue structure are needed to predict measures of patient progress.

Then, a large quantitative study (on texts) was conducted by the authors with interesting results in terms of conversation strategies that are associated with better outcomes; in particular, the most interesting aspects are: adaptability, dealing with ambiguity, creativity, making progress, and change in perspective.

We further describe some of these aspects as follows:

- *Reacting to ambiguity*: from the linguistic and the dialogic points of view, length, concreteness, strength of the reaction, and response style have been considered.

- To detect *temporal differences* in how counselors progress through different steps of the conversations and the protocol, the authors used a message-level Hidden Markov Model (HMM) to produce the distribution during each expectation step.

- In terms of *coordination and power differences*, the authors used eight linguistic coordination markers (as suggested by Danescu-Niculescu-Mizil [2012], and Danescu-Niculescu-Mizil et al. [2012]) to verify that the conversation partners adapt to each other's conversational style and that conversation participants who have a greater position of power coordinate less.

- *Perspective change* in the client over time is associated with higher likelihood of conversation success.

Finally, addressing depression, the authors proposed a novel measure to quantify three orthogonal aspects of perspective change within a single conversation: *time* (from issues in the past to the future), *self* (from talking about themselves to considering others and, potentially, the effects of their situations on others), *sentiment* (change in sentiment; i.e., the presence of a positive perspective change).

9.6.2 Framework and CA to Support Psychotherapist Training and Supervision: CORE-MI and ClientBot

Hirsch et al. [2018] noted that recent advances in ML and NLP provide effective methods to leverage spoken language, in psychotherapy sessions, as quality indicators and performance-based feedback [Pace et al., 2016]. Moreover, recent literature demonstrated that Motivational Interviewing (MI) sessions can be evaluated using ML and NLP, and that machine-coded sessions can be comparable with human-coded sessions [Atkins et al., 2014].

Imel et al. [2019] underline that ML algorithms can analyze audio and transcripts, generating rating of psychotherapy sessions that are consistent with traditional human-derived observer. Very few studies, however, face the usage of ML algorithms to give automatic feedback to psychotherapists.

The request of psychotherapists to be supported during their clinical activity, their training and supervision through AI-, NLP-, and ML-based frameworks and CAs is growing and has been considered by the research group involving (among others) Atkins et al. [2014], Hirsch et al. [2018], Imel et al. [2019], and Tanana et al. [2019].

The Counselor Observer Ratings Expert for Motivational Interviewing (CORE-MI by Hirsch et al. [2018]) is an interesting and innovative automated evaluation and assessment system that provides feedback to mental health counselors and psychotherapists on the quality of their activity.

The system has been evaluated by 21 counselors and trainees, considering the applicability of the system to clinical practice and the users' perception in terms of surveillance, workplace misuse, notions of objectivity, and system reliability.

The system has been built considering the MI methodology for clinical interventions, because of its effectiveness in promoting behavioral change, and its reliance on the therapeutic relationship (i.e., empathy and collaboration). The main goal of the system was to support the psychotherapist relational strategies, such as the use of open-ended questions, and making high-quality reflections of what the client said during the session.

The paper by Can et al. [2014] describes the processing steps and the goals of each step: vocal exchanges are segmented and assigned to the speakers; then, Automatic Speech Recognition (ASR) is applied to obtain transcriptions; and finally, text and speech predictive models are applied to obtain CORE-MI post-session reports. Additionally, paralinguistic information – such as prosody, pitch, speech rate, and intensity – are all estimated. A variety of ML approaches were used; in particular, the system makes use of the Barista open-source speech processing framework [Can et al., 2014].

CORE-MI provides feedback on standard MI quality measures, described in the Motivational Interviewing Treatment Integrity Scale [Moyers et al., 2010]. The report presents users with an overall *MI fidelity score*: a composite metric of the six standard summary measures of *MI quality* (empathy, MI spirit, reflection-to-question ratio, percent open questions, percent complex reflections, and percent *MI adherence*).

In particular, the two following evaluation concepts are quite interesting:

- *MI adherence* divides the total number of MI-adherent utterances (e.g., asking open questions, making complex reflections, supporting and affirming patients, and emphasizing client autonomy) by the sum of MI adherent and MI non-adherent counselor behaviors.

- *MI spirit and empathy* captures the "gestalt" of the session, assessing the overall competence of the counselor along the dimensions of: collaboration, evocation, and autonomy. Moreover, empathy measures the extent to which the counselor tries to understand the client's perspective.

Counselors in community clinical practice are rarely evaluated, and never by an automated system. To prevent disruptive reactions or not acceptance, the authors conducted a study of user attitudes towards automated evaluation. The following were formulated:

- *Receptivity*: how open are counselors to the concept of automated evaluation?

- *Workflow*: what role, if any, can counselors imagine automated evaluation playing in their clinical practice?

- *Concerns*: what concerns, if any, do counselors have about introducing automated evaluation into their practice?

The system usage was particularly appreciated by psychotherapists for two main reasons:

- Psychotherapists could compare scores from measures on the CORE-MI with their perceptions of how they conducted therapy (therapeutic style and experience level)

- CORE-MI could be integrated into the supervision and training of new counselors by generating detailed feedback for trainees and opening discussion of development of specific skills.

Another original use of CAs for psychotherapist training is described in Tanana et al. [2019]. The authors developed and evaluated ClientBot, a patient-like neural CA, which provides real-time feedback to trainees via chat-based interaction. NLP models were used by Tanana et al. to replicate behavioral coding evaluations of psychotherapy and to create the opportunity for simulating a standardized patient without the cost of recruiting and training human patients. According to Tanana:

> Training therapists is both expensive and time-consuming. Counseling skills practice often involves role-plays, standardized patients, or practice with real clients. Performance-based feedback is critical for skill development and expertise, but trainee therapists often receive minimal and subjective feedback, which is distal to their skill practice.

To address the challenges related to the need for scale and immediacy in training new skills in psychotherapy, the authors developed and evaluated a Web-based system that uses machine learning–based feedback for training 2 specific counseling skills: open questions and reflections. The feedback is embedded into a text-based neural conversational agent, developed to be a standardized patient. Thus, the skills training relied on an automated standardized patient—ClientBot—which provided real-time feedback to trainees on their utilization of specific counseling skills. (…)

The text-based conversational agent was trained on an archive of 2354 psychotherapy transcripts and provided specific feedback on the use of basic interviewing and counseling skills (i.e., open questions and reflections—summary statements of what a client has said). A total of 151 non-therapists were randomized to either (1) immediate feedback on their use of open questions and reflections during practice session with ClientBot or (2) initial education and encouragement on the skills.

From Tanana et al. [2019]

Satisfaction with the ClientBot system in general, and Satisfaction with the ClientBot Simulated Client were both measured: the majority of respondents said that system was not boring and that they thought the system gave useful information.

9.7 CONCLUSIONS AND FUTURE RESEARCH

A reasoned survey of the most recent literature describing CA, NLP, and ML solutions for psychotherapy and specialists' training was proposed with a double intent: to allow the enrichment of clinical interventions and to invite NLP and ML experts to consider this applicative domain as particularly important for the development of CAs, their interactive models, and their evaluation.

The research experiences of the author underline aspects that are still neglected and should be deeply analyzed and better modeled and implemented in new solutions:

- The vocal and prosodic dimension of dialogues is a powerful channel of affective and intention expression [Cenceschi et al., 2018; Schuller and Batliner, 2014; Schuller et al., 2013; Wennerstrom, 2001].

Paralinguistic feature, considered with linguistic ones, influence the reciprocal tuning [Rocco et al., 2018; Sbattella et al., 2014] and support empathic behaviors allowing the transformation of the self in psychotherapeutic relations.

- The multilevel nature of interactions should be taken into account to support a rich analysis of interactions in psychotherapy, for training and supervision (as in Sbattella et al., [2014] where an HMM-based framework was applied to analyze interrogations in law courts): emotions, interpersonal motivational systems [Liotti and Monticelli, 2008], contents and narrative, moments of discontinuity at different levels, and speakers' personalities should be considered and made evident to the trainees.

- Particular attention should be paid to the discontinuities: the specialist should be supported with instruments allowing the analysis (at different levels) of what happened before and after a discontinuity moment.

- Complex psychotherapeutic intervention (and not only micro-intervention) should be adequately analyzed and supported. More psychotherapeutic models should be addressed and implemented.

- Specific research should be dedicated to defining methodologies that allow to create specialized corpora of clinical dialogues and results, while respecting client privacy. New methodologies should be able to anonymize data to cope with the ML capabilities of understanding identities through the management of big and dispersed data.

- Recent models should be used to consider the client's and therapist's personalities. Not only language-based personality models (such as the Big5 by Costa and McCrae [1992]) should be considered, but also models that consider personalized ways of feeling, integrating and regulating – specifically and globally – body reactions, emotional skills, and cognitive aspects orienting dialogue, relations and transformative psychotherapeutic processes (such as Panksepp [2018]).

As outlined in this chapter, considering that one of the most important aspects of the psychotherapeutic interventions deals with empathic resonance and affective management, and that these capabilities are going to be better studied in the near future by NLP and ML, we have to encourage a clinical and technological research work that promises to be interesting, rich, and socially useful.

REFERENCES

Aafjes-van Doorn, K., Kamaeeg, C., Bate, J., Aafjes, M., *A scoping review of machine learning in psychotherapy research*, Psychotherapy Research, doi: 10.1080/10503307.2020.1808729, 2020.

Abd-Alrazaq, A. A., Alajlani, M., Alalwan, A. A., Bewick, B. M., Gardner, P., Househ, M., *An overview of the features of chatbots in mental health: A scoping review*, International Journal of Medical Informatics, doi: 10.1016/j.ijmedinf.2019.103978, 2019

Abd-Alrazaq, A., Alajlani, A., Alalwan, M., Denecke, A. A., Bewick, K., Househ, M., *Perceptions and opinions of patients about mental health chatbots: Scoping review*, Journal of Medical Internet Research, 23(1), p. e17828, doi: 10.2196/17828, 2021.

Abd-Alrazaq, A. A., Rababeh, A., Alajlani, M., Bewick, B., Househ, M., *Effectiveness and safety of using chatbots to improve mental health: Systematic review and meta-analysis*, Journal of Medical Internet, 22(7), p. e16021, doi: 10.2196/16021, 2020.

Abdulrahman, A., Richards, D., *Modelling working alliance using user-aware explainable embodied conversational agent for behaviour change: Framework and empirical evaluation*, in 40th International Conference on Information Systems, ICIS 2019, pp. 1–17, Munich, Germany, December, Association for Information Systems, 2019.

Adikari, A., De Silva, D., Alahakoon, D., Yu, X., *A cognitive model for emotion awareness in industrial chatbots*, in 2019 IEEE 17th International Conference Industrial Information INDIN, pp. 183–186, URL: http://dx.doi.org/10.1109/INDIN41052. 2019.8972196, 2019.

Adikari, A., De Silva, D., Moralyiage, H., Alahakoon, D., Wong, J., Gancarz, M., Chackochan, S., Park, B., Heo, R., Leung, Y., *Empathic conversational agents for real-time monitoring and co-facilitation of patient-centered healthcare*, Future Generation Computer Systems, 126, pp. 318–329, doi: 10.1016/j.future.2021.08.015, 2022.

Agarwal, S., LeFevre, A. E., Lee, J., et al., *Guidelines for reporting of health interventions using mobile phones: mobile health (misreporting and assessment (mERa) checklist*, BMJ, 352, p. i1174, 2016.

Althoff, T., Clark, K., Leskovec, J., *Large-scale analysis of counseling conversations: An application of natural language processing to mental health*, Transactions of Association for Computational Linguistics, 4, pp. 463–476, 2016.

Andersson, G., Cuijpers, P., Carlbring, P., Riper, H., Aedman, E., *Guided internet-based vs. face- to-face cognitive behavior therapy for psychiatric and somatic disorders: A systematic review and meta-analysis*, World Psychiatry, 13(3), pp. 288–95, 2014.

Ardito, R. B., Rabellino, D., *Therapeutic alliance and outcome of psychotherapy: Historical excursus, measures, and prospects for research*, Frontiers in Psychology, 2, p. 270, 2011.

Asada, M., *Towards artificial empathy. How can artificial empathy follow the developmental pathway of natural empathy?* International Journal of Social Robotics, 7, pp. 19–33, doi: 10.1007/s12369-014-0253-z, 2015.

Atkins, D. C., Rubin, T. N., Steyvers, M., Doeden, M. A., Baucom, B. R., Christensen, A., *Topic models: A novel method for modeling couple and family text data*, Journal of Family Psychology, 26(5), pp. 816–827, doi: 10.1037/a0029607, 2012.

Atkins, D. C., Steyvers, M., Imel, Z. E., Smyth, P. *Scaling up the evaluation of psychotherapy: Evaluating motivational interviewing fidelity via statistical text classification*, Implementation Science, 9(1), doi: 10.1186/1748-5908-9-49, 2014.

Balint, E., *The possibilities of patient-centered medicine*, The Journal of the Royal College of General Practitioners, 17(82), pp. 269–276, 1969.

Beatty, C., Malik, T., Meheli, S., Sinha, C., *Evaluationg the therapeutic Alliance with a free-text CBT conversational agent (Wysa): A mixed-methods study*, Frontiers in Digital Health, 4, doi: 10.3389/fdgth.2022.847991, 2022.

Beck, A. T., *Depression: Clinical, Experimental, and Theoretical Aspects*, University of Pennsylvania Press, Philadelphia, PA, 1967.

Bendig, E., Erb, B., Schulze-Thuesing, L., Baumeister, H., *The next generation: Chatbots in clinical psychology and psychotherapy to foster mental health – A scoping review*, Verhaltenstherapie, doi: 0.1159/000499492, 2019.

Berman, J. J., *Principles of Big Data: Preparing, Sharing, and Analyzing Complex Information*, Elsevier, Morgan Kaufmann, Waltham, MA, 2013.

Bickmore, T., Gruber, A., *Relational agents in clinical psychiatry*, Harvard Review of Psychiatry, 18(2), pp. 119–130, 2010.

Bickmore, T., Gruber, A., Picard, R., *Establishing the computer-patient working alliance in automated health behavior change interventions*, Patient Education and Counseling, 59(1), pp. 21–30, doi: 10.1016/j.pec.2004.09.008, 2005.

Bickmore, T., Mitchell, S., Jack, B., et al, *Response to a relational agent by hospital patients with depressive symptoms*, Interact Computer, 22(4), pp. 289–298, 2010

Bird, T., Mansell, W., Wright, J., Gaffney, H., Tai, S., *Manage your life online: A web-based randomized controlled trial evaluating the effectiveness of a problem-solving intervention in a student sample*, Behavioural and Cognitive Psychotherapy, 46(5), pp. 570–82, 2018.

Bossuyt, P. M., Reitsma, J. B., Bruns, D. E., *STARD 2015: An updated list of essential items for reporting diagnostic accuracy studies.* BMJ, 351, p. h5527, 2015

Burns, M. N., Begale, M., Duffecy, J., Gergle, D., Karr, C. J., Giangrande, E., Mohr, D. C., *Harnessing context sensing to develop a mobile intervention for depression*, Journal of Medical Internet Research, 13(3), pp. 158–174, doi: 10.2196/jmir.1838, 2011.

Can, D., Gibson, J., Vaz, C., Georgiou, P. G., Narayanan, S. S., *Barista: A framework for concurrent speech processing by USC-SAIL*, in 2014 IEEE International Conference on Acoustics, Speech and Signal Processing (ICASSP 2014), Florence, Italy, 4–9 May 2014 (Piscataway, NJ, May 2014), pp. 3306–3310, 2014.

Carlbring, P., Andersson, G., Cuijpers, P., Riper, H., Hedman-Lagerlöf, E., *Internet-based vs. face-to-face cognitive behavior therapy for psychiatric and somatic disorders: An updated systematic review and meta-Analysis*, Cognitive Behavioral Therapy, 47(1), pp. 1–18, 2018.

Cenceschi, S., Sbattella, L., Tedesco, R., *Towards automatic recognition of prosody*, in *9th International Conference on Speech Prosody 2018*, Poznan, Poland, 13–16 June 2018, pp. 319–323, 2018.

Chung, J., Gulcehre, C., Cho, K. H., Bengio, Y., *Empirical evaluation of gated recurrent neural networks on sequence modeling*. arxiv.org, (12 2014), http://arxiv.org/abs/1412.3555, 2014.

Cohen, S., Kamarck, T., Mermelstein, R., *Perceived stress scale. Measuring stress: A guide for health and social scientist*, Oxford University Press, New York City, USA, 1997.

Costa, P. T. Jr., McCrae, R. R., *Four ways five factors are basic*, Personality and Individual Differences, 13, pp. 653–665, 1992.

Craig, P., Dieppe, P., Macintyre, S., Michie, S., Nazareth, I., Petticrew, M., *Developing and evaluating complex interventions: The new medical research council guidance*, BMJ, 337, p. a1655, 2008.

Crutzen, R., Peters, G.-J. Y., Portugal, S. D., et al., *An artificially intelligent chat agent that answers adolescents' questions related to sex, drugs, and alcohol: An exploratory study*, Journal of Adolescent Health, 48 (5), pp. 514–519, 2011.

Damasio, A., *Self Comes to Mind. Constructing the Conscious Brain*, Pantheon, New York City, NY, 2010.

Damasio, A., *Feeling and Knowing. Making Mind Conscious*, Pantheon, New York City, NY, 2021.

Danescu-Niculescu-Mizil, C., *A computational approach to linguistic style coordination*. Ph.D. thesis, Cornell University, 2012.

Danescu-Niculescu-Mizil, C., Lee, L., Pang, B., Kleinberg, J., *Echoes of power: Language effects and power differences in social interaction*, WWW, 2012.

Des Jarlais, D. C., Lyles, C., Crepaz, N., *Improving the reporting quality of non-randomized evaluations of behavioral and public health interventions: The TREND statement*, American Journal of Public Health, 94(3), pp. 361–366, 2004.

Diener, E., Wirtz, D., Biswas-Diener, R., Tov, W., Kim-Prieto, C., Choi, D., Oishi, S., *New measures of well-being. Assessing well-being*, pp. 247–266, 2009.

Diener, E. D., Emmons, R. A., Larsen, R. J., Griffin, S., *The satisfaction with life scale*, Journal of Personality Assessment, 49(1), pp. 71–75, 1985.

Dwyer, D. B., Falkai, P., Koutsouleris, N., *Machine learning approaches for clinical psychology and psychiatry*, Annual Review of Clinical Psychology, 14(1), pp. 91–118, doi: 10.1146/annurev-clinpsy-032816-045037, 2018.

Eysenbach, G., *Consort-ehealth group, Consort-ehealth: Improving and standardizing evaluation reports of web-based and Mobile health interventions*, Journal of Medical Internet Research, 13(4), p. e126, 2011.

Fiske, A., Henningsen, P., Buyx, A., *Your robot therapist will see you now: Ethical implications of embodied artificial intelligence in psychiatry, psychology, and psychotherapy*, Journal of Medical Internet Research, 21(5), p. e13216, doi: 10.2196/13216, 2019.

Fitzpatrick, K. K., Darcy, A., Vierhile, M., *Delivering cognitive behavior therapy to young adults with symptoms of depression and anxiety using a fully automated conversational agent (Woebot): A randomized controlled trial*, JMIR Mental Health, 4(2), p. e19, 2017.

Fulmer, R., Joerin, A., Gentile, B., et al., *Using psychological artificial intelligence (Tess) to relieve symptoms of depression and anxiety: Randomized controlled trial*, JMIR Mental Health, 5, p. e64, 2018.

Gardiner, P. M., McCue, K. D., Negash, L. M., Cheng, T., White, L. F., Yinusa-Nyahkoon, L., et al., *Engaging women with an embodied conversational agent to deliver mindfulness and lifestyle recommendations: A feasibility randomized control trial*, Patient Education and Counseling, 100(9), pp. 1720–9, 2017.

Graham, S., Depp, C., Lee, E. E., Nebeker, C., Tu, X., Kim, H.-C., Jeste, D. V., *Artificial intelligence for mental health and mental illnesses: An overview*, Current Psychiatry Reports, 21(11), p. 116, doi: 10.1007/s11920-019-1094-0, 2019.

Hirsch, T., Soma, C., Mered, K., Kuo, P., Dembe, A., Caperton, D., Atkins, D. C., Imel, Z. E., *"It's hard to argue with a computer:" Investigating psychotherapists' Attitudes towards automated evaluation*, in Proceedings of the 2018 Designing Interactive Systems Conference (DIS '18), Association for Computing Machinery, New York, NY, USA, pp. 559–571, 2018.

Horvath, A. O., Greenberg, L. S., *Development and validation of the working Alliance inventory*, Journal of Counseling Psychology, 36(2), pp. 223–233, doi: 10.1037/0022-0167.36.2.223, 1989.

Howes, C., Purver, M., McCabe, R., *Linguistic Indicators of Severity and Progress in Online Text-Based Therapy for Depression*, in *Proceedings of the* Workshop *on* Computational Linguistics and Clinical Psychology: From Linguistic Signal to Clinical Reality, *Association for Computational Linguistics*, Baltimore, MD, pp. 7–16, 2014.

Hutto, C. J., Gilbert, E., VADER: *A parsimonious rule-based model for sentiment analysis of social media text*, in Proceedings of the Eighth International AAAI Conference on Weblogs and Social Media, AAAI Publications, Ann Arbor, MI, pp. 216–225, 2014.

Imel, Z. E., Coperton, D. D., Tanana, M., Atkins, D. C., *Technology-enhanced human interaction in psychotherapy*, Journal of Counseling Psychology, 64(4), pp. 385–393, doi: 10.1037/cou0000213, 2017.

Imel, Z. E., Pace, B. T., Soma, C. S., Tanana, M., Hirsch, T., Gibson, J., Georgiou, P., Narayanan, S., Atkins, D. C., *Design feasibility of an automated, machine-learning based feedback system for motivational interviewing*, Psychotherapy (Chic), doi: 10.1037/pst0000221, 2019.

Inkster, B., Sarda, S., Subramanian, V., *An empathy-driven, conversational artificial intelligence agent (Wysa) for digital mental well-being: Real-world data evaluation mixed-methods study*, JMRI Mhealth and Uhealth, 6(11), p. e12106, https://mhealth-jmir.org/2018/11/e12106, 2018.

Ireland, D., Atay, C., Liddle, J., *Hello Harlie: Enabling speech monitoring through chat-bot conversations*, Studies in Health Technology and Informatics, 227, pp. 55–60, 2016.

Joerin, A., Rauws, M., Ackerman, M. L., *Psychological artificial intelligence service, Tess: Delivering on-demand support to patients and their caregivers: Technical report*, Cureus, 11, pp. 1–7, 2019.

Krause, K., Guertler, D., Moehring, A., Batra, A., Eck, S., Rumpf, H.-J., Bischof, G., Lucht, M., Freyer-Adam, J., Ulbricht, S., John, U., Meyer, C., *Feasibility and acceptability of an intervention providing computer-generated tailored feed- back to target alcohol consumption and depressive symptoms in proactively recruited health care patients and reactively recruited media volunteers: Results of a pilot study*, European Addiction Research, 25(3), pp. 119–131, doi: 10.1159/000499040, 2019.

Laranjo, L., Dunn, A. G., Tong, H. L., Baki Kocaballi, A., Chen, J., Bashir, R., Surian, D., Gallego, B., Magrabi, F., Lau, A., Coiera, E., *Conversational agents in healthcare: A systematic review*, Journal of the American Medical Informatics Association, 25(9), pp. 1248–1258, 2018.

Liotti, G., Monticelli, F., *I sistemi motivazionali nel dialogo clinico*, Raffaello Cortina Editore, Milano, 2008.

Lucas, G. M., Rizzo, A., Gratch, J., *Reporting mental health symptoms: Breaking down barriers to care with virtual human interviewers*, Frontiers in Robotics and AI, 4, pp. 1–9, 2017.

Ly, K. H., Ly, A., Andersson, G., *A fully automated conversational agent for promoting mental well-being: A pilot RCT using mixed methods*, Internet Interventions, 10(2017), pp. 39–46, 2017.

Mauriello, M. L., Tantivasadakarn, N., Mora-Mendoza, M. A., Lincoln, E. T., Hon, G., Nowruzi, P., Simon, D., Hansen, L., Goenawan, N. H., Kim, J., Gowda, N., Jurafsky, D., Paredes, P. E., *A suite of Mobile conversational agents for daily stress management (Popbots): Mixed methods exploration study*, JMIR Formative Research, 5(9), p. e25294, 2021.

Meadows, R., Hine, C., Suddaby, E., *Conversational agents and the making of mental health recovery*, Digital Health, 6, doi: 10.1177/2055207620966170, 2020.

Miner, A., Chow, A., Adler, S., Zaitsev, I., Tero, P., Darcy, A., Paepcke, A., *Conversational agents and mental health: Theory-informed assessment of language and affect*, in Proceedings of the Fourth International Conference on Human Agent Interaction, doi: 10.1145/2974804.2974820, 2016b.

Miner, A. S., Milstein, A., Schueller, S., *Smartphone-based conversational agents and responses to questions about mental health, interpersonal violence, and physical health*, JAMA Internal Medicine, 176(5), pp. 619–25, 2016a.

Miner, A. S., Shah, N., Bullock, K. D., Arnow, B. A., Bailenson, J., Hancock, J., *Key considerations for incorporating conversational AI in psychotherapy*, Frontiers in Psychiatry, 10 doi: 10.3389/fpsyt.2019.00746, 2019.

Morris, R. R., Kouddous, K., Kshirsagar, R., Schueller, S. M., *Towards an artificially empathic conversational agent for mental health applications: System design and user perceptions*, Journal of Medical Internet Research, 20(6), p. e10148, 2018.

Moyers, T. B., Martin, T., Manuel, J. K., Miller, W. R., Ernst, D., *Revised Global Scales: Motivational Interviewing Treatment Integrity*, 3.1.1 (MITI 3.1.1), University of New Mexico Center on Alcoholism, Substance Abuse and Addictions (CASAA), Albuquerque, New Mexico, USA, 2010.

Pace, B., Tanana, M., Xiao, B., Dembe, A., Ba, C., Soma, C., Steyvers, M., Narayanan, S., Atkins, D., Imel, Z., *What about the words? Natural language processing in psychotherapy*, Psychotherapy Bulletin, 51(1), pp. 17–18, 2016.

Panksepp, J., Biven, L., *The Archeology of Mind*, W. W. Norton & Company, Inc., New York, NY, 2012.

Panksepp, J., Davis, K. L., *The Emotional Foundation of Personality. A Neurobiological and Evolutionary Approach*, W. W. Norton & Company, Inc., New York, NY, 2018.

Paredes, P. E., Gilad-Bachrach, R., Czerwinski, M., Roseway, A., Rowan, K., Hernandez, J., *PopTherapy: Coping with stress through pop-culture*, in Proc. 8th International Conference on Pervasive Computing Technologies for Healthcare, Oldenburg, Germany, doi: 10.4108/icst.pervasivehealth.2014.255070, 2014.

Pennebaker, J. W., Mehl, M. R., Niederhoffer, K. G., *Psychological aspects of natural language use: Our words, our selves*, Annual Review of Psychology, 54(1), pp. 547–577, 2003.

Pestian, J. P., Matykiewicz, P., Linn-Gust, M., South, B., Uzuner, O., Wiebe, J., Cohen, K. B., Hurdle, J., Brew, C., *Sentiment analysis of suicide notes: A shared task*, Biomedical Informatics Insights, 5(Suppl. 1), pp. 3–16, 2012.

Philip, P., Bioulac, S., Sauteraud, A., et al. *Could a virtual human be used to explore excessive daytime sleepiness in patients?* Presence Teleoperators Virtual Environ, 23(4), pp. 369–376, 2014.

Philip, P., Micoulaud-Franchi, J.-A., Sagaspe, P., et al. *Virtual human as a new diagnostic tool, a proof of concept study in the field of major depressive disorders*, Scientific Reports, 7, p. 42656, 2017.

Picard, R., *Affective Computing*, Massachusetts Institute of Technology Press, Cambridge, Massachusetts, USA, 1997.

Provoost, S., Lau, H. M., Ruwaard, J., Riper, H., *Embodied conversational agents in clinical psychology: a scoping review*, Journal of Medical Internet Research, 19(5), p. e151, 2017.

Ramirez-Esparza, N., Chung, C., Kacewicz, E., Pennebaker, J. W., *The psychology of word use in depression forums in English and in Spanish: Testing two text analytic approaches*. In ICWSM, 2008.

Richards, D., Caldwell, P., *Improving health outcomes sooner rather than later via an interactive website and virtual specialist*, IEEE Journal of Biomedical and Health Informatics, 22(5), pp. 1699–1706, 2017.

Rocco, D., Pastore, M., Gennaro, A., Salvatore, S., Cozzolino, M., Scorza, M., *Beyond verbal behavior: An empirical analysis of speech rates in psychotherapy sessions*, Frontiers in Psychology, 9, p. 978, 2018.

Rubeis, G., Steger, F., *Internet- und mobil-gestützter interventionen bei psychischen störungen. Implementierung in Deutschland aus ethischer sicht*, Nervenarzt, 90(5), pp. 497–502, 2019.

Salman, S., Richards, D., Caldwell, P., *Analysis and empathic dialogue in actual doctor-patient calls and implications for design of embodied conversational agents*, IJCOL [online], Italian Journal of Computational Linguistics, 7–1, 2, pp. 91–112, 2021.

Sbattella, L., Tedesco, R., Trivilini, A., *Forensic examinations: Computational analysis and information extraction*, in Proc, of the 2nd International Conference on Forensic Science - Criminalistic Research (FSCR), pp. 31–40, Global Science Technology Forum (GSTF), Singapore, November, 2014.

Scholten, M. R., Kelders, S. M., Van Gemert-Pijnen, J. E., *Sel-guided web-based interventions: Scoping review on user needs and the potential of embodied conversational agents to address them*, Journal of Medical Internet Research, 19(11), p. e383, 2017.

Schuller, B. W., Batliner, A. M., *Computational Paralinguistics. Emotion, Affect and Personality in Speech and Language Processing*, John Wiley & Sons, Ltd, Chichester, West Sussex, UK, 2014.

Schuller, B. W., Steidl, S., Batliner, A., Burkhardt, F., Devillers, L., Müller, C., Narayanan, S., *Paralinguistics in speech and language. State-of-the-art and the challenge*, Computer Speech and Language, 27, pp. 4–39, doi: 10.1016/ j.csl.2012.02.005, 2013

Scotti, V., *Unsupervised Hierarchical Model for Deep Empathetic Conversational Agents*, pages 53–73. CRC Press, Boca Raton, 2023b. ISBN 978-1-003- 29612-6. DOI 10.1201/9781003296126-5. URL https://doi.org/10.1201/ 9781003296126-5.

Scotti, V., Galati, F., Sbattella, L., Tedesco, R., *Combining deep and unsupervised features for multilingual speech emotion recognition*, in Proc. Part II of Pattern Recognition, ICPR International Workshops and Challenges, Milan, Italy (held online), 2021.

Scotti, V., Sbattella, L., Tedesco, R., *A modular architecture for empathetic conversational agents*, in Proc. of IEEE International Conference on Big Data and Smart Computing, BigComp 2021, Jeju Island, South Korea (held online), 2021

Shatte, A. B. R., Hutchinson, D., Teague, M., *Machine learning in mental health: A scoping review of methods and applications*, Psychological Medicine, 49(9), pp. 1426–1448, doi: 10.1017/S0033291719000151, 2019.

Stieger, M., Nißen, M., Rüegger, D., Kowatsch, T., Flückiger, C., Allemand, M., *PEACH, a smartphone- and conversational agent-based coaching intervention for intentional personality change: Study protocol of a randomized, waitlist controlled trial*, BMC Psychology, 6(1), p. 43, 2018.

Suganuma, S., Sakamoto, D., Shimoyama, H., *An embodied conversational agent for unguided internet-based cognitive behavior therapy in preventative mental health: Feasibility and acceptability pilot trial*, JMIR Mental Health, 5(3), p. e10454, 2018.

Tai, A. M. Y., Albuquerque, A., Carmona, N. E., Subramanieapillai, M., Cha, D. S., Sheko, M., Lee, Y., Mansur, R., McIntyre, R. S., *Machine learning and big data: Implications for disease modeling and therapeutic discovery in psychiatry*, Artificial Intelligence in Medicine, 99, p. 101704, doi: 10.1016/j.artmed.2019.101704, 2019.

Tanana, M. J., Soma, C. S., Srikumar, V., Atkins, D. C., Imel, Z. E., *Development and evaluation of ClientBot: Patient-like conversational agent to train basic counseling skills*, Journal of Medical Internet Research, 21(7), p. e12529, doi: 10.2196/12529, 2019.

Vaidyam, A. N., Linggonegoro, D., Torous, J., *Changes to the psychiatric chatbot landscape: A systematic review of conversational agents in serious mental illness*, The Canadian Journal of Psychiatry, 64(7), pp. 456–464, 2019.

Watts, B. V., Shiner, B., Zubkoff, L., Carpenter-Song, E., Ronconi, J. M., Coldwell, C. M., *Implementation of evidence-based psychotherapies for posttraumatic stress disorder in VA specialty clinics*, Psychiatric Services, 65(5), pp. 648–653, doi: 10.1176/appi.ps.201300176, 2014.

Wennerstrom, A., *The Music of Everyday Speech. Prosody and Discourse Analysis*, Oxford University Press, 2001.

Yalçin, Ö. N., *Modeling Empathy in Embodied Conversational Agents*, Extended Abstract, Boulder, CO, USA, ICMI, doi: 10.1145/3242969.3264977, 2018.

Index